Debugging the Link between Social Theory and Social Insects

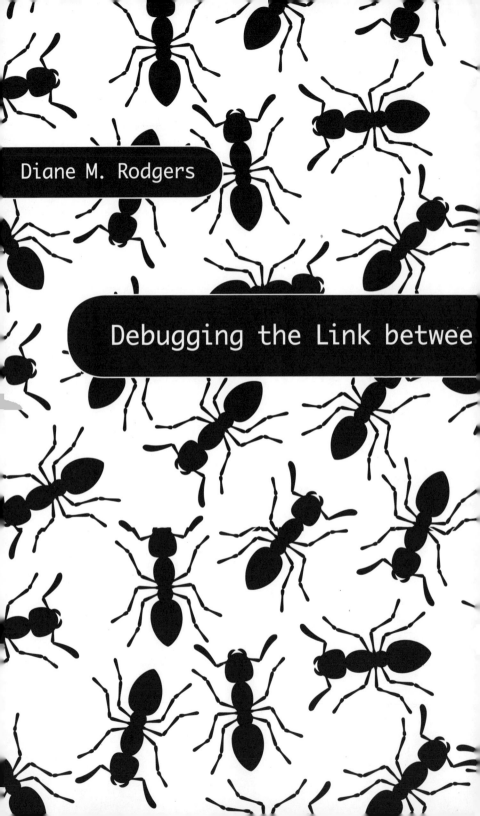

Diane M. Rodgers

Debugging the Link betwee

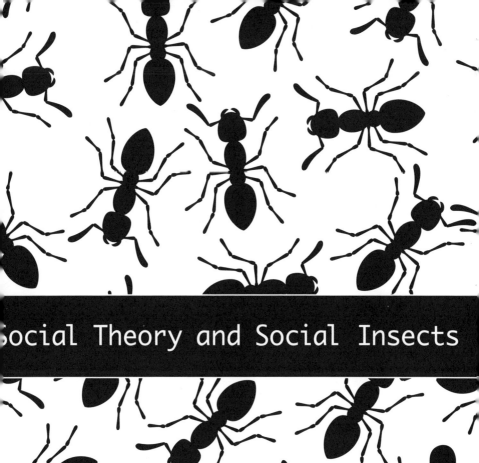

Social Theory and Social Insects

Louisiana State University Press

Baton Rouge

Published by Louisiana State University Press
Copyright © 2008 by Louisiana State University Press
All rights reserved
Manufactured in the United States of America

LSU Press Paperback Original
FIRST PRINTING

Designer: AMANDA MCDONALD SCALLAN
Typeface: WHITMAN, RD MONACO
Printer and binder: THOMSON-SHORE, INC.

Library of Congress Cataloging-in-Publication Data
Rodgers, Diane M., 1959–
  Debugging the link between social theory and social insects / Diane M. Rodgers.
    p. cm.
  Includes bibliographical references and index.
  ISBN 978-0-8071-3369-9 (pbk. : alk. paper) 1. Interpersonal relations. 2. Social structure. 3. Insect societies. 4. Sociology. 5. Entomology. I. Title.
  HM1106.R634 2008
  156—dc22

2008017784

The paper in this book meets the guidelines for permanence and durability of the Committee on Production Guidelines for Book Longevity of the Council on Library Resources. ∞

Contents

Acknowledgments vii

Introduction 1

1. Insect Sociality and Why It Matters for Human Sociality 4

2. The Need for a Critical Approach 20

3. A Bee or Not a Bee: *Historical and Cross-Cultural Interpretations* 40

4. Entomologists and Sociologists: *Comon Ground* 63

5. Despite the Differences: *Insect Sociality and Comparative Method* 91

6. Naturalizing Hierarchical Sociality through Discourse 115

7. Alternative Visions of Insect and (Human) Sociality 155

Conclusion 181

References 189

Index 205

Acknowledgments

I would like to sincerely thank the following people for sharing their expertise with me during the development of the ideas in this book: Ken Benson, Joan Hermsen, Tola Pearce, Wayne Brekhus, James Carrel, David N. Smith, and Mohan K. Wali. Their thoughtful suggestions and comments on the work enhanced it greatly. Much appreciation goes out to Althea Harris and Morgan Matsiga for bravely reading the earliest drafts.

I am grateful to Joseph B. Powell, Wesley Shrum, and Louisiana State University Press for their enthusiasm about publishing this book. Editor Joseph Powell has made this an exciting and interesting experience, offering invaluable guidance and cheerfully answering my many questions. I am extremely appreciative of the insightful remarks given by the anonymous reviewer, which strengthened the work. And many thanks to Susan Brady, copy editor extraordinaire.

For graciously helping me with my last-minute request for technical assistance and graphics troubleshooting, I am indebted to Mariano Spizzirri and Robert Banke. Stacy James contributed the original design of the hierarchy triangle. Alexandra Chapman and Megan Sprangers rescued me by helping with manuscript preparation.

Thanks to my friends/academic colleagues who offered needed advice, encouragement, and hours of conversation along the way: Matt Lammers, Theresa Goedeke, Carrie Prentice, Kathy Doisy, Mary Jo Neitz, La Tanya Skiffer, Soo Yeon Cho, Tola Pearce, Veronica Medina, and Nancy Turner Myers. To all my other dear friends who were so kind and understanding during this long process, I could not have finished this book without your support, companionship, and patience. And, as always, unending love and gratitude to MBF.

Debugging the Link between Social Theory and Social Insects

Joan,
Thanks so much for all the encouragement & guidance!
Diane Rodgers

Introduction

Social insects such as ants, bees, and wasps have captured the human imagination and fostered numerous analogies to human social structure. In addition to being popular in the general public discourse, these analogies have generated theory and concepts within scientific literature. Specifically within the disciplines of entomology and sociology, the theory and concepts that liken social insect organizational structure to human organizational structure share an interesting history. A particular classification scale for social insects and the terminology that describes their behavior sprang from the interaction of the two disciplines during the nineteenth and early twentieth centuries. The depiction of the organizational structure of social insects has been co-constructed through an interchange of ideas that reflects not only representations of natural phenomena, but also specifically human social and political concerns.

This depiction of social insects has varied based on the historical time period, political persuasion, and cultural location of its human creators. Although variations in these depictions and their interpretations exist, the discipline of entomology has adopted analogies and classification schemes that resemble Western bureaucratic organizational structures, particularly in the tenets of a specialized division of labor and hierarchy. Socially constructed hierarchies of race, class, and gender are found in the lexicon and classification of social insects; the entire premise of sociality for insects is predicated on a dominant understanding of high/low civilization standards and comparisons that appear to be informed by nineteenth-century colonial thought. Within the history of sociology, these same concepts and analogies were used to "naturalize" hierarchical human social structures. This historical interaction became a legitimating loop between the emerging disciplines of sociology and entomology, borrowing credibility from both the social and natural worlds.

Because of this correlation and the naturalization of these sociological theories and concepts, I approach the historical entomological and sociological literature using a theoretical framework that deconstructs these generally accepted ideas about sociality and social organization. This critical science theoretical framework is composed of post-Kuhnian, feminist, and postcolonial science studies. Given the type of hierarchical analogies found in the scientific discourse, each of these approaches has strengths to offer separately and combined within a critical science studies theoretical framework. Because naturalized accounts of the social structure tend to deemphasize the details of their historical and social creation, it is necessary to recover these origins through a critical discourse analysis. Fairclough (1995) describes this recovery process as one of "denaturalization." A critical discourse analysis of the key entomological and sociological texts of the nineteenth and early twentieth century reveals the co-construction of analogies between social insects and humans that reinforced an interlocking hierarchical social structure.

One of my goals in this work is to give readers unfamiliar with entomology a better understanding of the field by providing sufficient background and contextualizing this information through historical and cross-cultural evidence. To this end, I present an in-depth explanation of the definition and terms used in the research on and discussion of social insects, including the insect sociality scale. Scales of sociality are by no means confined to entomology; since its inception, sociology used scales of sociality that generally reflected colonial viewpoints about civilized and primitive societies. I provide an overview of these nineteenth-century social evolutionary scales and their correlation to entomology. The generally accepted classification schemes, models, and terms in the dominant discourse are specifically Western and therefore situated in a Western worldview.

There have been previous analyses of social insect symbolism, analogies, and classification, but not from a critical science studies perspective. The use of that perspective in the analysis presented in this book will complement the growing body of entomological literature that is challenging the traditional understanding of insect sociality and social organization. As the entomologist Deborah Gordon frames the emerging issue, "We need new ways of understanding the organization of social insect colonies" (Gordon 1999, 67). Because analogies between social insects and humans are co-constructed, the reformulation of social organizational forms for humans is also a focus of this critical analysis. Understanding how human hierarchical structures were naturalized through comparison to natural models sheds light on future perceptions and changes in human social structures.

The hierarchical analogies that were developed naturalized race, class, and gender hierarchies in social organizational structure. There were also challenges to this construction within entomology and sociology. A closer look at the literature provides an opportunity to locate oppositional discourse, as both of these fields were initially more inclusive before rigid disciplinary boundaries came to be established. Oppositional discourse from the nineteenth and early twentieth centuries frames the current alternative models within this larger historical context. A specific account of interactions between entomology and sociology provides a closer look at how concepts and theories of sociality and social organization are co-constructed. This is valuable information for those involved in the ongoing co-constructions that shape our ideas about the social and natural world.

Alternative models are currently being proposed that challenge the existing definitions and terms used to discuss sociality and organizational structure in both sociology and entomology. Understanding the past co-constructions, how they were socially constructed, and the worldviews or standpoints they reflect is important for attempting to analyze current co-constructions. The organicism that informed many biological metaphors in the nineteenth and early twentieth century has experienced a resurgence in current analogies involving the intersection of computer systems, human groups, and insect societies. The perceived universality and interdisciplinary exchange of ideas in the creation of these models is reminiscent of earlier co-constructions surrounding the idea of sociality and social organization. The historical background and contemporary implications of this scientific discourse is of interdisciplinary interest.

Although current alternative models may appear to be a shift away from hierarchical analogies, in many cases the legacy of the historical co-constructions carries over into the new paradigms. There are serious implications of this for future interdisciplinary model making, especially as new models combine ideas about social insect and human organizational structure with computer technologies. In as much as alternative models retain the race, class, and gender hierarchies and the underlying assumptions of past terminology, they will serve to reinforce hierarchical social institutions. A concerted effort must be made to change the social insect lexicon and to "debug" this theoretical co-construction.

1

Insect Sociality and Why It Matters for Human Sociality

One of the most direct and lasting influences of the interaction between sociology and entomology may be found in the entomological definitions for social insects. Much group behavior can be deemed social in some sense, so why are certain insects particularly defined by this term? There are guidelines in entomology that determine the sociality of an insect and the level of sociality it exhibits. This is accomplished with a hierarchical classification scheme, describing those with the highest form of sociality as "eusocial" and the lower forms as "subsocial." These two general categories are subdivided to contain more subtle distinctions of more or less social behavior.

Within the category of eusocial insects fall ants, termites, and some species of bees and wasps. These are the insects that will be of primary interest in this book as they are the insects that are almost exclusively used in social theories. Currently in the field of entomology a very basic definition of the characteristics needed for the highest level of insect sociality includes three traits: (1) division of labor, especially a reproductive division of labor with some sterile individuals; (2) caring for the young by the colony; and (3) overlap of generations. This definition is a compilation of previous definitions (Hermann 1979). In 1928, William Morton Wheeler, a prominent entomologist specializing in ants, proposed the first systematic definition for eusocial insects that focused on caring for the young and overlap of generations. Later definitions placed more emphasis on the division of labor as a requirement for eusociality, a trend beginning with Charles Michener and Mary H. Michener (1951) and E. O. Wilson (1971). The term "eusocial" was introduced in 1966 by

Suzanne W. T. Batra (1966) and then employed by Wilson in his definition, which remains the dominant one used in the literature.

The category of eusocial is further divided between primitive eusocial insects and more specialized eusocial insects. Basically, the more specialized the division of labor, the more eusociality will be ascribed to an insect. Evidence of physical dimorphism is seen as characteristic of higher specialization. If a specialized division of labor with a caste system is a marker for eusociality, more communal behavior falls into the subsocial category. Those defined as social insects make up the smallest percentage of the population of insects, while the other less social and solitary insects comprise the greatest part. A way of visualizing this relationship of scale and populations would be in a descending triangle (fig. 1).

Is there a hierarchy within the category of eusocial insects itself? Although all species of termites and ants are eusocial, bees and wasps have a limited number of eusocial species. However, this difference does not seem to be a major factor in considering one type of eusocial insect to be more social than the others. There still appears to be some informal ranking within the category of eusociality, albeit one that may be subject to change over time and that may depend on any given researcher's particular leanings.

Ants and bees seem to be the most widely studied and discussed of all the social insects. Honeybees have a long, singular history of being stud-

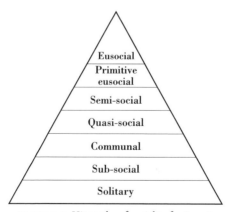

FIGURE 1. Hierarchy of sociality for insects

ied due to their practical uses in human society, such as the production of honey and pollination of crops. Beginning as early as the seventeenth century, the behavior and social structure of honeybees have been extensively described by entomologists. For instance, hailed as "the first work of literature in the world dealing with insects, thus finally establishing entomology, and especially systematical entomology, as a science," *De animalibus insectis libri VII*, published by Ulysse Aldrovandi in 1602, provided a detailed description of the honeybee (Beier 1973, 85).

Two of the most renowned entomologists in the twentieth and twenty-first centuries are William Morton Wheeler and E. O. Wilson, both known for their research on ants. Wilson, along with his co-author Bert Hölldobler, won the Pulitzer Prize for *The Ants* in 1991, and Wheeler's 1910 book *Ants: Their Structure, Development and Behavior* is considered a classic entomological text. A comment of Wheeler's indicates that scientists may vie for positioning one species of eusocial insect over another, especially in regard to the two most widely studied—ants and bees: "The problem of the sex of the offspring of parthenogenetic female insects has been much befogged by the eternal discussion, especially in Germany, of the honey-bee, an exceptional creature which it is very difficult to dislodge from zoölogical text-books and lecture rooms as the paragon of social insects" (1928a, 162). Wheeler's statement seems to express concern about using honeybees as a model to explain all eusocial behavior. His concern over bees being viewed as the ideal social insect may have had to do with his wanting to give ants that title, in part because they were his area of specialization.

The issue of which insects provide the best model of eusociality can be illustrated by studies with the wasp family Vespidae, specifically the paper wasp genus *Polistes*. Not all wasps are eusocial, and therefore primitively eusocial wasps are seen as a model for learning about the evolution of eusociality in wasps (West-Eberhard 1996). However, they may not be seen as the best model to study the organization of eusociality itself as compared to ants and bees. Although eusocial wasps are being studied more frequently than in the past, especially the *Polistes* species, one reason they are less likely to easily fit as a model for the eusocial category is that their colonies have more than one queen. Another issue may be that in some of these wasp species there is much more physical similarity between workers and queens, making it hard to tell them apart, unlike ant castes. Both of these differences tend to blur the boundaries for some of the theories and definitions of eusociality. The question of what the best model might be, therefore, often leads to the question: the best model for *what* (Burian 1996)?

Termites serve as an interesting model of eusocial insects in that they are considered in some regards to be the "perfect" eusocial insect because of their highly specialized division of labor, with distinct physical differences between types of workers, as well as their having both a queen and king as the primary reproductives. However, research on termites does not have the same extensive history as that on either bees or ants. In fact, termites were initially confused with ants, and were called "white ants." Wheeler's explanation for why they were not as widely studied as other eusocial insects is a curious one: "Owing to the fact that they are nearly all tropical insects and do not make attractive cabinet specimens, they were much neglected by entomologists till the beginning of the present century" (1928a, 135). This may not be a plausible reason for the neglect of termites; many insects not deemed "attractive" do elicit interest. Termites certainly were a part of nineteenth-century natural history collections that included tropical specimens. One early major work on termites, *The Soul of the White Ant,* was written by Eugène N. Marais and published in 1937 in South Africa (the work appeared even earlier, beginning in 1923 as a series of articles). Marais' work did not get much attention outside of South Africa until what some claimed to be a plagiarized version of it appeared in Maurice Maeterlinck's book *The Life of the White Ant* in 1926 (Ardrey 1969).

There are various reasons that a given eusocial insect might become more predominant than others in the literature. An entomologist who becomes well known may cause his research specialty to become equally well known. A particular species may serve as a better model for dominant theories concerning social insects or sociality during any given time period. Access to and ease of observation of the insect also may play a role, as may the practical use of the insect for any given culture.

CONTENTION OVER DEFINITION AND TERMS

The criteria for classifying levels of sociality and the terminology used to describe social insect behavior are sometimes questioned by entomologists and have been significantly challenged within the discipline of entomology (Costa 1997, 2006; Costa and Fitzgerald 2005; H. Evans 1984; Foster 2002; Gordon 1989, 1999; Milius 2000; Reznikova 2003; Sherman et al. 1995; Wcislo 1997, 2005; West-Eberhard 2003). The main contention is that certain species other than ants, bees, wasps, and termites have qualities that should place them within the eusocial category. One such case involves several species of aphids, which have now begun to earn the title of eusocial. This "prized" designation is being won through a gradual process of persuasive evidence and eventual integration. Aphids had been considered subsocial until the discovery of a soldier caste prompted some researchers to push for recategorization of several species. In addition, when researchers found that the aphid species *Pemphigus obesinymphae* exhibited conflict over reproduction, especially through the soldier caste, Foster (2002, 199) claimed that, "It is perhaps time to admit them ungrudgingly to the elite cadre of the truly social insects." Although this statement might suggest the decision had not yet been made as to whether this aphid species could be termed eusocial, Foster began his article with an affirmative statement that gave the impression that aphids had already been admitted into the elite category: "Aphids hold a special place in the study of social evolution, because they include the only eusocial animals that are clonal" (2002, 199). Another reframing occurred when Foster added a "queen" to the specialized division of labor already exhibited by the accepted soldier caste (2002, 199). This bid to include aphids in the eusocial category made a case for inclusion by asserting that aphids adhered to the eusocial criteria, not by questioning the criteria. Other insects, such as thrips and beetles, have also undergone this process of induction by entomologists desiring their acceptance as eusocial. The exclusionary nature of the classification system is in these cases being questioned by those specializing in other groups of insects and even some species of spiders. James T. Costa's *The Other Insect Societies* (2006) catalogs and describes these marginalized insect groups and attempts to stake out some level of social status for their behavior and organizational structure.

Less often are critiques leveled that question the entire hierarchy of the sociality scale. It is on this important issue that I will focus, rather than arguing for the elevation of certain species to the eusociality category. Attempts to classify more insects as eusocial serve to reify the category. My intention is rather to question the adoption of this category, the scale of sociality, and the terms of sociality. Certain definitions of and terms regarding insect sociality are commonly accepted and utilized within contemporary Western science. The transfer of ideas that emerge in the terms or scale of sociality is therefore rooted in modern Western culture. In deeming certain insects as "social," it may be difficult not to interpret the meaning of "social" in human terms. Human understanding of the social is situated in particular cultures and time periods, reflecting a social construction of ideas about social insects. What is needed, therefore, is a sociological look at how sociality is described in the entomological literature, and specifically how it has been co-constructed with social theory since the nineteenth century.

A certain amount of anthropomorphism seems to accompany the discussion of social insects, and this may be becoming more prevalent, as Kennedy notes of the "new anthropomorphism" (1992). Ants and termites are currently described as "stealthy invaders" and "reserve labourers"; they develop "mass transit systems" and are involved in "arms races"; bee colonies have "effective policing"; social insects are capable of "spite" and "corrupt motives" (Boswell et al. 2001; T. Evans 2006; Franks 2001; Gardner and West 2004; Heinze et al. 2006; Ratnieks and Wenseleers 2005). New terms are developed, and the older anthropomorphic descriptors continue to be used. This anthropomorphism is accepted because it is viewed as simply a convenient way of describing behavior. However, it also creates sociobiological analogies that then contribute to the naturalization of social theories and concepts. The anthropomorphism shapes a particular view of nature through a social lens and naturalizes social structure through a reinforcing loop.

The terminology used for the roles within social insect colonies (workers, soldiers, and queens) is clearly influenced by a social interpretation of the natural world. Many of these terms were created in the nineteenth and early twentieth centuries but currently remain in the classification system. Occasionally they are put in quotation marks or otherwise qualified in published research by authors who then proceed to use the terms throughout. The basic terms found in Western accounts are:

Queen. The queen is seen as responsible for the primary reproductive role in the colony of all social insects. She has also been considered the one who coordinates all activities of the workers, although more recently this role has been questioned altogether or described as having varying levels of influence. Some eusocial insects such as wasps have more than one queen.

King. The male reproductive of a colony. The role of the king is noted only with termite colonies because he remains living with the queen.

Drone. This term especially refers to male bees; they exist only to mate with the queen, dying or being killed off after that activity.

Worker. A female nonreproductive of a colony who performs a wide variety of tasks including maintaining the nest or hive, gathering food, and taking care of the young. In termites, sterile males are also called workers.

Soldier. A worker who usually has specialized physical features and who defends the colony.

Slave. Within ant colonies, an ant of a species other than the host that was taken from its original colony to be raised in the host species' colony and performs the role of worker there. "Auxiliary worker" is an alternative term used to describe this type of ant.

Slave-maker. An ant that takes ants from another colony to raise within their own colony as workers. Sometimes this behavior is termed "dulosis" and these ants are termed "dulotic." Ants exhibiting this behavior in the genus *Polyergus* are sometimes called "Amazons."

Guard bees. These bees stay at the entrance of the hive to ward off intruders. They detect by scent if an incoming bee is part of the colony. Guard bees that are near the entrance but actually attack intruders are sometimes referred to as soldier bees.

Nurse. A nonreproductive worker who tends to the eggs laid by the queen.

Scout. A worker who leaves the colony to forage for food, needed materials for the colony, or a new site if the nest is to be moved.

Farmer. An ant that cultivates fungus and stores it for food within their colony. These cultivations are then called "fungus gardens."

Dairyman. An ant that uses aphids for nourishment, drinking in the honeydew that is produced by the ant stroking the aphid. Aphids are termed "cows" in this situation.

Army ant. This term is reserved for particular species of ants, mostly the subfamilies Dorylinae and Ecitoninae, also referred to as legionary ants. Army ants are nomadic and prey as a group on other species, including mammals. A soldier ant is not a synonymous term for an army ant.

Thief/robber. These ants are so small that they can invade another species nest easily and take from their stock of food.

The terms for roles that are used are also correlated to larger anthropomorphized organizational activities, physical structures, and institutions. The behavior and activities of social insects are said to encompass: marriage, armies and army maneuvers, slave raids, robbery, and altruism. These activities are described as taking place in cities, villages, factories, the royal chamber, and bivouacs. The insect colonies are perceived to be overseen by all types of governments (socialistic, democratic, republican, anarchic, and nepotistic). This elaborate anthropomorphism makes eusocial insect colonies especially conducive to being utilized as analogues for human society.

A FEW MORE TERMS AND HOW THEY WORK TOGETHER

Eusocial insect colonies have been termed a "superorganism" because each individual is so interdependent with the group that the group itself takes on the characteristic of an organism. This term comes from Herbert Spencer, who applied it to human societies but also felt that it could be applied to describe a universal condition for any complex organism. The term had fallen out of favor because of an increase in studying the colony based on individual selection pressures, but it has recently been revived as interconnectedness becomes recognized in both new technical systems, such as the Internet, and more general reinterpretations of the natural world.

Within the division of labor in a colony, there are specialized roles that result in adult forms or types that are considered "castes" and "subcastes." The diversity of castes varies depending on the particular eusocial insect; for instance, termites have a king, but ants, bees, and wasps do not. A soldier subcaste appears in worker ants, bees, and termites, yet with termites the soldier subcaste is extremely distinct in physical form as compared to those in that group within ant or bee colonies. When castes become so physically different, this is considered to be a case of polymorphism, whereas the fact that castes or subcastes perform different behavioral tasks is termed polytheism. In some eusocial insects, these tasks result in age-related castes and therefore are not a permanent or physically altering specialization (H. Evans 1984, 204).

Why do workers perform tasks in the colony at all? What has been considered the "problem" of altruism has been an ongoing question within the literature on eusocial insects, beginning with Darwin. One of the most influential answers to this problem has centered on the kin-relatedness of the workers to each other. This can be explained by the reproductive system of haplodiploidy, whereby only females emerge from fertilized eggs and are thus termed diploid; males are considered haploid because they emerge from unfertilized eggs. It is evolutionarily more advantageous for females to work with their sisters to whom they are most closely related genetically than to have offspring of their own. This creates what is known as a female worker caste. Haplodiploidy is the key feature for the kin selection theory of William Hamilton, which was published in the early 1960s and became the predominant explanation for altruism until recent challenges of it on many fronts (Freeman and Herron 2004). As mentioned earlier, there are disagreements among entomologists over terminology, as well as over specific ideas concerning the division of labor, caste, and polymorphism. In comparison, there is general acceptance of the definition of eusociality, which is characterized by an overlap of generations, parental care, and a division of labor. Each of these characteristics, therefore, deserves some further description.

THE THREE FACTORS IN THE DEFINITION OF EUSOCIALITY

Overlap of Generations

Although it is one of the three basic requirements for being considered eusocial, the overlap of two or more generations is a trait shown in many insects. The overlap of generations is assumed to occur within the same nest for social insects. Aphids, a group of insects not considered eusocial with the exception of one species, exhibit a remarkable case for the overlap of generations. They are viviparous, that is, capable of live birth and producing fully formed offspring in utero. However, almost all the species of aphids seem to lack the other two accepted characteristics of eusociality: parental care and division of labor. Another reason aphids may not qualify to be defined as eusocial is because of the nature of their reproduction. As Michener and Michener (1951, 2) describe aphids and others: "First and least social are those strangely self-sufficient insects whose females are able to produce a family without mating, without so much as seeing, feeling or smelling another individual of another species. This type of reproduction is called parthenogenesis." Because of the additional requirement of parental care, overlap of generations alone will not qualify aphids for eusociality; some semblance of family life must occur to fulfill the criteria of the definition.

Parental Care

First noted by Wheeler as a requirement for a higher level of sociality, parental care corresponds to an emphasis on the family as the basic unit of analysis for evolutionary development. In his scale of sociality, Wheeler used parental care as the key to dividing up social insects into further levels of sociality: "There has evidently been a very long evolution through numerous stages of constantly increasing intimacy of the mother with her progeny from the most rudimentary stage of complete or almost complete indifference to one of mutual and abiding cooperation" (1928a, 12).

Wheeler numbers the progressive stages of "motherhood" from one to seven. Each stage falls into a category that is accorded importance relative to its level of sociality. He terms categories one through five "infrasocial" and category six as "quasi-" or "subsocial." While he does not use the term "eusocial," he does indicate that, unlike the others, the "final stage" is fully social. In this stage, "[t]he progeny are not only protected and fed by the mother, but eventually co-operate with her in rearing additional broods of young, so that parent and offspring live together in an annual or perennial society" (1928a, 13). The parents and offspring in this highest social stage live together due to an "increased interest on the part of the mother in the later instars of her offspring," which appears to be helped along by her increased life span (1928a, 13). In 1951, Charles Michener and Mary Michener also provided a scale of parental care similar to Wheeler's and claimed: "We can measure the social development of an insect species in terms of the care which the mothers expend on their offspring. We find a progression from what we might call shameful neglect to the most attentive rearing in these general groups" (2). Although parental care and overlap imply a reproductive division of labor, which had been discussed in earlier definitions, current usage places even more emphasis on division of labor as a defining quality of eusociality.

Division of Labor

E. O. Wilson placed more emphasis on caste and the reproductive division of labor in his definition of eusociality. In a recent interview, Wilson described the three criteria for eusociality: "First, there are two major castes—a queen, or sometimes a king, which constitutes a reproductive caste, and workers that don't reproduce as much if at all. Second, you have generations of grown, mature adults living with other grown mature individuals in the same community. And finally, you have mature adults that take care of the young. Those three elements are the primary criteria of what makes an advanced insect colony" (2003, 3). While parental care remains important, Wilson and Hölldobler consider the caste system the "defining property of eusociality" (Cromie 2005, par. 13). Division of labor as a central concept for both social insect and human organization relies on ideas from the social sciences. The entomologist Deborah Gordon notes that Adam Smith's concept of the division of labor for human society was influential in the development of ideas about the behavior of eusocial insects. She takes issue with this application and especially with the way that Oster and Wilson (1978) depicted the importance of this type of specialization for social insects. She further contends that insect "task-switching" is a much more prevalent behavior than a division of labor that emulates any human model (1989). It seems clear that the three criteria for insect eusociality have been influenced by social theory. This influence becomes a mutual and reinforcing factor for both sociology and entomology.

CONNECTIONS TO SOCIAL THEORY AND THE HUMAN SCALE OF SOCIALITY

What is the significance for sociological theory of the definition of a social insect and the accompanying terms? Early sociological theories adhered to ideas about social evolution that incorporated scales of sociality for humans. They also used natural models to legitimate the formulations of social theories. Biological analogies were common in sociology throughout the nineteenth and early twentieth centuries, and social insect societies served prominently as natural models to compare to human societies. The development of definitions and terms used in these analogies can be seen in part as a co-construction by sociology and entomology. It is therefore important to trace the historical parallels as well as any contemporary ones that still exist or may be revived. Many of the terms used to describe insect sociality have their analogues in human society. Some, such as "queen," "king," "worker," "soldier," "farming," "marriage," and "armies," are readily apparent analogous terms. Less obvious, however, is how the terms "eusocial," "subsocial," "quasi-social," and "communal," and the hierarchical scale of sociality in general, mirror a colonial vision of the ranked order of cultures. The scale of sociality for insects is comparable to the social evolutionary scales that were used for humans during the nineteenth century and through the mid-twentieth century.

An example of this can be found in Franklin Henry Giddings's 1901 classification of human sociality, which parallels the hierarchical category of sociality in insects. He creates a scale of sociality classes and distinguishes between social and unsocial. As Giddings tells us: "Not all men associate habitually with the same individuals, or associate with any individuals in the same degree. Association therefore develops the social nature of different individuals in different degrees. It more or less fits them to be satisfactory and useful members of the community" (1901, 261). Giddings explains the differentiation in role status as the result of the individual's ability to adapt their social nature to society. Those who are unable to manage this are ranked as unsocial and those capable of this are social; these then become specific groups or classes. "These different reactions produce in the population inequalities and gradations of social nature. They create sociality classes" (1901, 261). Giddings's category of the "social" is further broken down into subcategories of low, medium, and high. Some of the qualities of the high sociality class are similar to the characteristics used to designate insects as eusocial. "*The High Sociality Class* is composed of those in whom the social nature is developed in the highest degree. They are not only socialized, but also *individualized* and distinguished. They not only participate in general, and especially in altruistic, organization, but they also plan and direct it" (1901, 262). In a description reminiscent of the scale for eusocial insects, Giddings labeled highly social groups as specialized, exhibiting altruistic tendencies but also displaying leadership and organization.

Spencer also used the idea of leadership and division of labor to construct a scale of simple, compound, doubly compound, and trebly compound societies that categorized every imaginable society from the Tasmanians to the "Teutons before the 5th century" and "England after the 10th century" (Carneiro 1967, 50–51). Every human society was positioned on this scale of simple to complex in Spencer's model. Spencer utilized anthropological data in his social scale, and these data were heavily influenced by the nineteenth-century worldview of colonialism. In fact, many sociologists and entomologists wrote volumes exclusively on the topic of social evolution from a colonial viewpoint alongside their other sociological and entomological texts. The contents of these works demonstrate a borrowing of concepts, as the overarching laws were seen to be the same.

The work of Sir John Lubbock illustrates the range of topic typical in the nineteenth century. A lack of rigid disciplinary boundaries and an expectation that scientific pursuits would be broad allowed for investigation into many arenas. Lubbock was highly esteemed, not only as an entomologist, but also as a politician serving in Parliament (J. Clark 1997). He was author of a wide variety of scientific works, but he also wrote moralistic essays. In the back of his 1882 book *Ants, Bees and Wasps*, a classic entomological text, the publisher lists some of Lubbock's other titles. They range from anthropological work such as *The Origin of Civilization and the Primitive Conditions of Man* to *Addresses, Political and Educational*, and then entomological work, including such titles as *On the Origin and Metamorphoses of Insects*. The diverse arenas of Lubbock's expertise tended to inform his writing as a whole. Lubbock applied the laws of social evolution equally to humans and insects and drew sociopolitical conclusions from his studies as well. The examples of ideas regarding evolution and hierarchy from the works of Giddings, Spencer, and Lubbock are typical of social theory during the late nineteenth century.

The ancient idea of a Great Chain of Being—an earlier worldview that included animals, humans, and God alike in a fixed ranking system—also influenced the development of the evolutionary hierarchy of sociality. One of the important changes that occurred during the shift in worldviews was eliminating the fixed nature of the system, which altered the boundaries between rankings. Evolutionary ideas had begun to take hold, promoting a new paradigm that suggested progress and change as crucial in the formation of the universe. As Lovejoy describes it, the idea of the Great Chain of Being became temporalized in the eighteenth century: "the entire created universe . . . came to be explicitly conceived, no longer as complete once for all and everlastingly the same in the kinds of its components, but as gradually evolving from a less to a greater degree of fullness and excellence" (1936, 316).

With the Great Chain of Being temporalized by the eighteenth century, the foundation was laid for the introduction of evolution in the nineteenth. Hierarchy itself was temporalized rather than refuted altogether, and it therefore became a part of the concept of evolution. Anthropological and biological data were presented in a fashion that highlighted forms that evolved from the simple to complex. During the same time period, this repackaged hierarchical idea proved useful in justifying the colonization of other people, as their cultures could be seen as being less evolved than those of the colonizers and as being lower within the social hierarchy. Comparisons between various human social groups as well as between human groups and eusocial insects were frequently made in both the sociological and biological literature. As specific criteria were being established for what constituted differences between primitive and civilized in a way that corresponded to the power structure of colonialism, a "social Darwinism" was being created. What is less clear is that social scales were also being grafted onto the interpretation and definition of natural scales of sociality. Tracing the co-construction of the definition and terms for eusociality in insects illustrates this mutual influence. Because this history still affects how social and natural systems are being conceived of today, the legacy of hierarchical ranking deserves exploration. As Juliet Clutton-Brock (1999, 19) writes concerning this legacy: "Despite all the changes that have taken place in biological science over the last 100 years, most people still believe that the world is ordered according to a hierarchical Scale of Nature with unicellular organisms at its base and Man at the top. This is not surprising since, from the moment of birth, people in the Western world are ruled by hierarchies, first in the family, then in education, and on through adulthood."

Because of internal disagreements among entomologists over the current definitions and terms for eusocial insects, I am optimistic that the idea of insect sociality can be reconfigured without the strong bias toward hierarchy. Theories of natural and social systems were co-constructed during the nineteenth and early twentieth centuries, and this seems to be reoccurring. Examining our understanding of how eusocial insects are used in co-construction is therefore relevant in analyzing the choice of social theories that are used to explain human social structure.

The definitions and terms used for social insects should be of concern to sociologists as well as entomologists. There is a history of co-construction surrounding hierarchical sociality scales within the entomological and sociological literature. With the advent of new technologies, our ideas concerning both natural and social systems are currently in flux, and the reformulations of theories, definitions, and terms involve new collaborations and co-constructions. On one front, the definition and terms used to describe social insects are being challenged, yet all the while some of the already existing terms are becoming further entrenched in the literature. A trend toward universal application is becoming more widespread, and new ideas such as self-organizing models or superorganisms are being viewed as representations of the natural, technical, and social world. Just as earlier social theories and models were socially co-constructed, so are these new models and paradigms. A critical look at the historical co-construction of terms describing social insects and the origins of the sociality scale is imperative to better understand the contemporary use of analogies and terms that naturalize social structure.

2

The Need for a Critical Approach

> Any perceived pattern or organized system in nature is liable to be employed to express and comment upon social order and social experience. In being so employed, the perceived pattern is itself liable to be developed and reconstituted to better fit it to its functions. The pattern in question may be of many kinds: the overall order of the cosmos, the system of natural kinds of plants and animals, the general organization of the earth's crust, even the humdrum routines of the honeybee
> —Barry Barnes and Steven Shapin, eds., *Natural Order*

SCIENCE, NATURE, AND THE SOCIAL: ISSUES OF AUTHORITY

How does the authority of modern science, the natural world, and the social world merge or overlap to legitimate each other? The scientific disciplines of entomology and sociology are situated within the larger natural and social realms of authority; therefore this larger context must be examined. A critical theoretical approach helps to explain the "legitimating loop" between the natural and the social, and the interaction that reinforces authority for both arenas.

Certainly, if we take the concept of nature as the starting point for legitimation, we see advantages to endowing all things social with the credibility given to the natural world. In particular, when "nature" is seen as a fixed object for study, it becomes a serious legitimating force. As Harvey points out, "The advantage in seeing values as residing in nature is that it provides an immediate sense of ontological security and permanence" (1996, 157). The work of early naturalists utilized the idea of observing and recording "objective nature" and most often mixed this with references to the current authority of religion. The sense that nature was outside the social, not to mention connected to a divine source, provided a reference for moral and civic lessons that otherwise might not hold as much authoritative weight (D. Allen 2004; Burke 1997). Once divine authority was conferred on nature, the door was opened for nature to take on its own independent status. With the assistance of "objective" human observers, natural laws were then derived from nature. The natural world became rational and could be used as a legitimating force for modern scientific institutions and theory. As Wright describes the transition: "In the world of scientific knowledge new kinds of institutions had to be legitimated and new theories of social life had to be developed, institutions and theories that derived their credibility from the rational idea of nature, not from the traditional idea of God" (1992, 118). A science built on the study of a socially created, rational nature became the legitimate source of knowledge for modern Western society.

From the seventeenth through the early twentieth century, the association of Western social thought with advances in the natural sciences began to add legitimation to social theory and especially to sociology as a discipline (Martindale 1960; Maus 1962). One of the ways that social theory attempted to gain this legitimation was through the use of analogies involving the natural world, which the natural sciences presented as rational. At this point there developed a legitimating loop for social theory and institutions whenever natural and social worlds were compared. In identifying how institutions achieve stability, Mary Douglas maintains that social classifications are naturalized. This is accomplished through finding an analogy that connects the social realm to an area outside itself. Therefore, "[w]hen the analogy is applied back and forth from one set of social relations to another and from these back to nature, its recurring formal structure becomes easily recognized and endowed with self-validating truth" (1986, 48). The social gains authoritative weight through the use of fixed, natural categories that then become embedded in social institutions, the structure of organizations, and the discourse. Wright argues that the scientific and political roots of social theory have to be consciously recovered because "a fundamental reference to objective nature, through the autonomous individual, necessarily underlies the development of both scientific social institutions and scientific social theory" (1992, 121).

At the same time, it is important to note that the idea of nature is defined by the social and is thereby subject to change over time and in relation to place. As such, it reflects social history and ideas about nature and is not a pure reflection of nature itself. In the interchange of analogies, how much has social theory influenced the understanding of the natural world? In the classification scheme for social insects, we can see the interconnections between theories about the social world and the natural world. In particular, we see the influence of the social on what is otherwise deemed an accurate reflection of the natural world.

Tracing the social construction of nature illustrates the subjective and fluid quality of these interpretations. However, applying critical theory adds the dimension of power dynamics to these constructions, taking the argument beyond complete relativism. Viewing the relationship between the social and the natural as dialectical shows the classification scheme for social insects to be as much about power dynamics as it is a reflection of the cultural norms of any place or time period. Scientific theories and language do not cause social or natural reality but rather become co-constructions in social structure and interpretations of the natural world that then create a loop of legitimation for ideas or institutions that are created or reinforced.

Hess (1995, 21) describes this same process as "technototemism," which includes the realm of the technical along with the natural in relation to the social. He derives this term from a basic definition of totemism as "the process by which social groups achieve coherence and distinctiveness by being identified with natural phenomena." Hess then applies this definition of totemism in a wider sense, making it useful for explaining the legitimation loop between the social and the natural or technical. Social and natural/technical categories reinforce meanings assigned to them by virtue of their connection with each other (1995, 22). Hess notes that some critical theorists, such as Engels, had previously determined that social relations could be inscribed onto nature and then back again to legitimate particular social relations as natural. Hess himself calls this the "boomerang of technototemism" but recognizes that other ways to describe this within critical theory would be termed "reification" or "mystification" (1995, 24).

Along with technototemism or reification, the idea that nature is being used to shape or justify the social is sometimes described simply as social Darwinism. However, the problem is not that laws of nature—in particular, evolutionary theory and natural selection—become attributed to the human social structure in a biased way that fosters inequality. A similar case of naturalization was used to justify cooperation and more egalitarian dynamics by those who disagreed with the emphasis on an individualistic, "survival of the fittest" social Darwinism (Cronin 1991; Hofstadter 1944; Sapp 1994). More to the point, the ideas of natural selection and evolution posited as natural laws cannot be seen apart from the social world or the social actors who had these ideas. Although the term "social Darwinism" implies that there could conversely be a Darwinism that is separate from the social, the connection to the social is inescapable. In using Spencer's ideas of the survival of the fittest and Malthus's theory of human population, Darwinism was created out of previous conceptions of the human social world: "[I]t seems that Darwin and Wallace found in Malthusian theory not social factors masquerading as 'natural' but a social theory that they had to 'naturalize'" (Cronin 1991, 273).

Once nature becomes perceived in a certain way and is then described in social terms, justification of the social can happen through comparisons, even though this is not a direct, unilateral comparison to some objective "natural reality." Nature is always perceived and described through its association with the social. A term such as "boomerang" or "co-construction" may accurately portray the relationship between the natural and social realms as mutual exchange, but the term "legitimation loop" provides this sense of mutual influence as well as implying the potential benefit this connection would provide both arenas. Therefore, I will use the term "legitimation loop" most consistently throughout this book.

SCIENCE STUDIES AND CRITICAL THEORY

Post-Kuhnian science and technology studies have long accepted the existence of a co-construction or a legitimation loop between the natural and the social, but have now turned to explicating the more specific cases and mechanisms that create these circumstances of mutual influence. One area within the science and technology studies literature that especially engages with this type of research question is the work focused on boundaries, classification, and issues involving knowledge production. This literature also tends to intersect with the sociology of knowledge and critical theory.

Up until the publication of Thomas Kuhn's *The Structure of Scientific Revolutions* in 1962, the history, philosophy, and sociology of science maintained an internalist approach to scientific knowledge claims. Science was seen as somehow above any social, political, or cultural influences, and therefore the examinations of scientific knowledge focused on areas such as "discoveries," "famous men," and "the scientific revolution in the West." When Kuhn opened the door to the possibility that external factors were involved in the development of scientific paradigms, science studies assumed a more critical tone (Restivo 1995; Sismondo 2004). Latour and Woolgar's *Laboratory Life* (1979) focused on the actual practice of science in laboratories and included a critical assessment of the terminology that scientists used to describe their work. The technical language of science was analyzed as a social production rather than being assumed to be a part of objective scientific activity. Other critical studies on the practice of science and the use of scientific terminology followed Latour and Woolgar's publication (Gilbert and Mulkay 1984; Knorr-Cetina 1981; Lynch 1982; Traweek 1988), and included a call for the explication of the political aspects of the production of scientific knowledge (Young 1985). It was thought that this would expose the legitimation of the social order through naturalization, opening the door for change. Taking account of the external factors involved in the production of scientific knowledge became the "new" sociology of science. Post-Kuhnian science studies offered the critical approach necessary to denaturalize scientific knowledge and scientific disciplinary formation.

SCIENCE STUDIES AND CLASSIFICATION

Science and technology studies (STS) have begun to engage in a critical assessment of classification systems. STS and contributions from other disciplines diverge from the previous internalist history of classification by critically examining the subjective qualities and social consequences of classification systems (Bowker and Star 1999; Dean 1979; Lakoff 1987; Ritvo 1997a; Roth 2005). All of these studies ask what purpose classification systems serve and by what processes they are enacted. These are questions that had traditionally not been asked. Roth explains that, "[c]lassification-in-use and classifying activity are ordinarily unproblematic and invisible—it is only when classification breaks down, when there is trouble of some form, that the nature and consequences of classification become apparent (2005, 583). The categorization of social insects calls for an attempt to answer some of these questions, both in the areas that have already begun to be challenged and in some that have not. It is true that disciplinary disagreements concerning the specifics of insect sociality and their classification can problematize the generally accepted, invisible categories. However, they can also be consciously problematized; the process and consequences of categorizing and classifying can be denaturalized and made clearer by using critical approaches to evaluate classification systems.

Classification occurs with the help of categories, and the origins of these categories have long been a source of debate. The hierarchical categories found in the concept of the Great Chain of Being were seen as a reflection of divine order; in the eighteenth century, as this set of rigid categories became temporalized, nature, rather than the divine order, came to be seen as the origin of the categories (Lovejoy 1936). Due to this paradigm shift, the development of categories for classification systems came to be viewed as a cognitive process that simply reflected the categories that existed in nature. This idea has come under critique in many disciplines. Foucault (1973) linked the ordering of the natural and social world to a shared interdisciplinary process that attempted to name an objective reality; he therefore proscribed the need for an "archaeology" to sift below the surface to understand this process of ordering. Lakoff (1987) documented a change in analyzing the process of categorization within the field of cognitive science as one transitioning from an idea of "objectivism" to "experientialism." In other words, he described a shift from believing that categories are an exact correspondence to external reality regardless of the one categorizing, to the understanding that categories are embodied and engage the human imagination (Lakoff 1987, xii–xv). With this shift to a new way of thinking about categories also comes "changes in the concepts of truth, knowledge, meaning, rationality—even grammar" (1987, 9). Lakoff claims that the previous views surrounding these concepts will be left behind in the wake of a new understanding of categorization. This new understanding would thereby eradicate all the underlying assumptions that accompany the classical theory of categories: Cartesian mind/body split, objective reality, universal truth, and the belief in a direct connection between symbol and meaning (1987, 9).

The earlier notion of classification embraced principles based on a modern conception of rationality and progress. Classification became legitimated as an objective science. The critique of the ideological underpinnings of the modern era has forged a more skeptical inquiry into areas previously regarded as objective and above sociopolitical influence. New approaches within linguistics take a constructionist and critical attitude toward classification and categories. Because language is not objective, neither are the classification systems that language is used to create. The way that any given society uses language to classify and categorize can be seen to reveal its ideological belief systems (Birch 1989, 168).

This new way of looking at categorization and the creation of classification systems in various disciplines has sparked new perspectives within science studies. Case studies on how classification systems are constructed reveal the vagaries of categorization. Making distinctions between some classification schemes over others is ideologically based and part of an ongoing and often continuous process. This "boundary work" (Gieryn 1983) can be intra- or interdisciplinary and is of significant import for the individuals involved in the struggle over demarcation (Bowker and Star 1999; Dean 1979; Roth 2005). Dean (1979, 226) believes that classification systems should be viewed through an "invention model"; seen in this light, classification is invented out of sociopolitical interests. He presents a case study of botany classification and concludes: "Even in a case like this, without 'external' sociopolitical factors being particularly evident, a sociological approach is essential to an understanding of the development and distribution of different classifications of the natural world" (1979, 227).

A sociological approach is also needed to examine the legitimation loop created with the exchange of natural and social categories used in classification systems. The way that classifications systems are designed can reinforce certain interpretations of both the natural world and the social world. This process can be illustrated in the creation of the scientific classification of a hierarchical scale of sociality for social insects. Although classification systems can appear to separate the realms of interchange between the natural world and the social world, they are interwoven. Bowker and Star point out that the invisibility of this connection adds to the ongoing ability of classification to appear objective. When categories and classification mystify the link between the natural and the social, they perform the "ever local, ever partial work of making it appear that science describes nature (and nature alone) and that politics is about social power (and social power alone)" (1999, 46). Not only can classification be demystified and be understood to have a relationship to social theory, but also those theories combined with classification systems can be shown to wield power in tandem. Scientific description and classifications are created by social beings and thus create or reinforce social dynamics. The study of classification can be strengthened through additional theoretical insights from feminist and postcolonial science studies that incorporate social location and relational dynamics of race, class, and gender into their research.

FEMINIST SCIENCE STUDIES

Early feminist science studies were concerned with recovering the presence and influence of women in science (Alic 1986; Gornick 1983; Herzenberg 1986; Rossiter 1982; Schiebinger 1989). Recovering the histories of marginalized women in the sciences served to challenge ideas about science as an exclusively masculine endeavor. This was a part of the larger move in academia by feminists to revise history to reflect women's contributions, especially in the academic disciplines. It was also a step toward correcting the underrepresentation of women in certain disciplines, including science. Despite the allegedly neutral quality of science, Keller (1985; 1995) pointed to a pattern of "genderization" within the field of science (1985, 76). Her perspective built on the claim that women were being devalued and underrepresented in science and placed this in the larger context of the unspoken assumption of science itself as masculine. The question then followed: If scientific knowledge was to be seen as masculine, would women "do science differently" when given the opportunity? The perspective of "difference feminism" was thus introduced into the early inquiries about gender and scientific knowledge. Keller's exemplary work on Barbara McClintock, *A Feeling for the Organism* (1983), presented McClintock as approaching her research in a more holistic manner that was influenced by her social location as a woman. Later feminist science studies, in utilizing the idea of social location, have tended to move away from such an essentialist position (although recovering histories of women's contributions to science is still an important endeavor). Mayberry, Subramaniam, and Weasel (2001, 5–6) present a more complex view of feminist science studies that rejects the idea of binary, static relationships. Although knowledge comes from a location, it is "situated" within a multidimensional space of race, class, and gender. The application of a feminist perspective fluidly challenges the socially constructed boundary between nature and culture; knowledge and practice. Haraway explains the need to see natural science as situated knowledge, unique in its legitimating authority. This authority exerts power over everyday people's lives and therefore needs to be examined in its particular guise. "The detached eye of objective science is an ideological fiction, and a powerful one. But it is a fiction that hides— and is designed to hide—how the powerful discourses of the natural sciences really work" (Haraway 1989, 13).

In understanding the development of scientific knowledge concerning social insects, I find particularly useful the theoretical work that has emerged from the feminist science analysis of primate studies (Asquith 1996; Haraway 1989; Sommer 2000; Strum and Fedigan 2000). Like social insects, primates have been used as a model of comparison to human societies. Feminist science theorists and some natural scientists themselves challenged the dominant discourse that revolved around the presentation of a hierarchical, male-dominated, and aggressive social organization of primates as a universal model. Feminist science studies pointed to the subjective nature of the choice of species model, and questioned the universalizing discourse that was being used to naturalize human racial, gender, and class hierarchies. As Sommer observed: "The danger of projecting human preoccupation with classification and hierarchies onto animal societies, especially anthropoids, is at all times present. Hierarchies we impose on the non-human world reflect the social principle of 'othering' people, excluding them from one's own sphere because of their gender, race or social class" (2000, 32–33).

Because of the commitment to the intersection of race, class, and gender, some feminist science studies include a combination of critical theory and postcolonial theory. Although this hybrid theory provides the necessary tool for analyzing intersectionality, what are the specific contributions of postcolonial science studies?

POSTCOLONIAL SCIENCE STUDIES

Because the co-construction of the definitions and terms for eusocial insects emerged from the West during the nineteenth and early twentieth centuries, the social influence of colonialism can be found embedded in the co-construction. The recovery and legitimation of indigenous knowledge claims about insect classification illustrate how the dominant discourse is structured by specifically Western definitions and terms. Therefore the dominant discourse can be critiqued as being specifically founded on a colonial worldview.

Colonial practice included adopting local knowledge and/or practices and quite often claiming them as their own. When this did not happen, the knowledge and practices of the colonized were ignored or destroyed. As Itty Abraham elaborates, "in the nineteenth century, as the relationship of colonial domination to knowledge began to be reflexively conceptualized, the colonial state began to identify, and in some cases eliminate, native or local sources of knowledge that seemed threatening or difficult to comprehend through contemporary codes of scientific practice" (2000, 53). In my later discussion of the social construction of definitions and terms for eusocial insects, I include indigenous knowledge of social insects. These indigenous models and terms for social insects in many cases differ from those used in the dominant discourse, underscoring social location and the significance of worldview as factors in creating such models and terms. What I present is meant to support this larger theoretical claim, rather than to focus on this aspect of the postcolonial science studies agenda. The research on ethnoentomology is growing and is an area that will provide more challenges to the dominant model and terms used in the literature on social insects.

The recent work in postcolonial science studies that critiques Western scientific knowledge claims is more germane to my argument. Postcolonial criticism emerged out of the field of subaltern studies and the attempt to name the colonized "other" of Western imagination as a subject that had agency outside Western discourse. This became complicated since dominant discourse/practices and subaltern subjects invariably exist in relation to each other. Prakash traces the transition of postcolonial critique from subaltern as subject to critique of the dominant discourse of the West. The history of the colonized inevitably leads to the colonizer and the colonizer's knowledge claims. When that happens, "the weapon of critique must turn against Europe and the modes of knowledge it instituted" (1994, 1483). This shift initially focused on the discipline of history but then moved into other fields. What distinguished postcolonial critique from a postmodern or poststructuralist engagement with text was its use of the concept "other" to explain the dominant discourse founded on dichotomous difference. Postcolonial science studies turned this same critique onto the discipline of science in an effort to decenter Western scientific knowledge claims (Harding 1998; Nandy 1988; Prakash 1994).

Social theories that incorporated a colonial worldview are not only abstractions to be deconstructed, denaturalized, or decolonized as text; their implications for practice should also be understood. Racial categories and classification systems within social evolutionary scales allowed for inhumane treatment of the colonized, who were depicted as lower on the hierarchal scale than their European colonizers. The colonial worldview incorporated the exploitation and subjection of indigenous populations into a category of "other" in relation to the colonized (Gailey 1996; Gould 1981). These scales were framed in scientific discourse, with some groups presented as being less evolved than others on a scale of sociality. Zeitlen (1997, 197) explains that the "evidence" was gathered from various times and places and then presented in linear fashion. Stages from "savage" to "civilized" were demarcated as if they represented the historical evolutionary progression of the West. Zeitlen cites Nisbet in his assessment of how these evolutionary developments were falsely arranged and thus represented, "not a theory of the actual course of development of a single social entity, but rather a series of 'stills' as in a movie film" (1997, 197). In the nineteenth and early twentieth centuries, comparisons were made in the entomological and sociological literature between "lower" and "higher" societies, both human and animal. At times these comparisons were made interchangeably, reinforcing the idea that all scales of sociality were linked to a universal law of evolutionary progress. Using postcolonial science studies to analyze this discourse helps to explain how these particular hierarchical social scales were informed by a colonial worldview.

Social evolutionary theory as a construct that emerged from the practices of colonialism can be analyzed with postcolonial theory. This analysis can assist in better realizing that the scales of sociality developed in the nineteenth century were informed by a hierarchical, Western worldview where difference became a way to rank based on perceived levels of higher and lower status. Because the Western construction of terminology to describe eusocial insects originated in the nineteenth and early twentieth centuries, these ideas about social structure and organization were also influenced by colonialism. Postcolonial science studies can provide a way to decenter this Western construction and shed light on the colonial worldview that informed it.

THEORETICAL INTERSECTIONS

Robert M. Young (1985, 191) noted that tracing the ideological assumptions of modern science leads to an understanding that "the same structures support the modern rationalizations of industrial capitalism, colonialism, and imperialism." Science cannot be analyzed separate from the context in which it has been formed; there is no division between science and society. The sociology of science moved beyond an internalist history of science and then began to incorporate a critical perspective that questioned just the false divide that Young identified. Post-Kuhnian science studies, which name scientists as actors in the construction of scientific knowledge, interject agency into what might otherwise be a tendency of critical theory to overemphasize structural constraints. For instance, Latour's actor network theory presents a picture of science occurring in a dynamic and interconnected way, even subscribing agency to nonhuman actors. A further understanding of agency comes from feminist and postcolonial science studies, with their focus on social location and the complex intersection of race, class, and gender. Feminist theory's "situated knowledge" and the concept of "other" for postcolonial theorists not only imply a dialectic but become specific as to the intricacies of that dialectic. In these theories, the assumption of power imbalances within dominant ideology and practice still allows for agency to varying degrees; for most, the agency is perceived as an achieved state, not an essential one.

In the nineteenth century, ideas concerning social evolution dovetailed with the increasing belief in the existence of a biological basis for evolution. The overarching theme of early evolutionary theories was that of linear progress from simple to more complex societies or forms. Societies were often described as moving from "primitive" to "civilized" states, with the colonized seen as primitive and the colonizers as civilized. And hierarchical social rankings did not end with such designations for different cultures. Gender hierarchies also were named within cultures. For those cultures deemed civilized, men were ranked as the most civilized, with women and children vying for second-place ranking. The category of female was also portrayed as lower on the evolutionary scale due to "inherent" limitations. Both indigenous people and women have been viewed as "other" and have occupied a low position in hierarchical classification systems. The theoretical tools of feminist and postcolonial theory deconstruct the dominant discourse by naming the intersection of race, gender, class, and colonialism as they attempt to legitimate hierarchical structures.

My theoretical framework draws from the tradition of critical theory as well as post-Kuhnian, feminist, and postcolonial science studies. Taken together, this constellation of theories can help explain the significance of the connection between social theory and social insects as constructed to reflect a Western scientific worldview and as intersecting with race, class, and gender. The implication is that this can then be deconstructed and reconfigured to encompass alternative models or conceptions.

METHODOLOGICAL APPROACHES

Critical theory, post-Kuhnian, feminist, and postcolonial science studies share similar ontological and epistemological concerns. One of the methodological assumptions within these theoretical perspectives is the deconstruction or denaturalization of what is regarded as objective fact. An emphasis on the process of knowledge creation and/or ongoing scientific practice exposes any notion of fixity or privileged perspective. The qualitative method of critical discourse analysis was the most appropriate choice for this analysis of the primary texts from the fields of sociology and entomology. This type of discourse analysis exposes the power dynamics involved in the creation of text, thereby denaturalizing their content. Sociohistorical and social constructionist methods also enhance this critical analysis by locating the creation of knowledge in time and place. The primary documents I analyze are from the entomological and sociological literature, specifically books, journals, and correspondence that contained the intersection of social theory and social insects from 1800 to 1945. Books and articles that appeared in the search sample but fell outside of the time period boundary of the nineteenth and early twentieth centuries were sometimes used as secondary resource materials.

The primary sociological sources included the work of the major social theorists of the time period studied. The dominant discourse involving social insects as models appeared within major works of sociology. In my discourse analysis, I include key articles from early volumes of the *American Journal of Sociology* as well as classic texts. These include works from Max Weber, Herbert Spencer, George Herbert Mead, Emile Durkheim, Franklin Henry Giddings, Charlotte Perkins Gilman, Peter Kropotkin, and others considered influential in sociology during that time period, as well as those leaving behind a lasting legacy.

Within the entomological literature, I focus on the classic texts from the entomologists and naturalists of the same time period who studied and wrote about social insects. This list includes, but was not limited to, the work of Auguste Forel, Maurice Maeterlinck, François Huber, Jean-Henri Fabre, Karl von Frisch, John Lubbock, and William Morton Wheeler. The works of generalists such as Charles Darwin were also examined in reference to social insects. The label of naturalist applies to some of these writers, and their work may even bridge the genres between scientific and more popular or educational literature. This is inevitable given the gradual stages of discipline formation at this time. Some scientists wrote for both the discipline of entomology and wider audiences. For instance, because of the public's interest in apiculture, books by entomologists such as Anna B. Comstock presented scientific information that appealed to the beginner beekeeper. Books like these also contained current scientific ideas about the organizational structures of social insects.

My sources also included literature reflecting evolutionary and ecological organicism. The nineteenth and early twentieth centuries were a pivotal time in the development of the disciplines of social and natural science. Not only were disciplinary boundaries in flux, but so were the paradigms that informed them. As the eighteenth century brought an end to the idea of a predetermined Chain of Being, nineteenth-century conceptions based on more evolutionary scales ushered in a significant transition to holistic thought concerning the social and natural patterns of living organisms. In essence, the early nineteenth century saw a shift from a mechanistic worldview to a more organic paradigm. The effect of organicism on the sciences combined with the impact of the evolutionary models of Buffon, Lamarck, Spencer, and Darwin added considerably to the biological and social scientific literature. This organicist paradigm held sway until the mid-1940s (Levine 1995; Taylor 1988), and facilitated many biological analogies, including social insect societies as models for human society. Theodor Eimer was one of the more influential of these organicist theorists. Eimer was especially noted for promoting orthogenesis, the idea of an internal, unilinear drive mechanism for evolution that rivaled natural selection. In his book *Organic Evolution,* "Eimer interpreted his organicism as entailing the notion that if the individual did not struggle for the good of it all, it would only do injury to itself. Taking as a model the social life of bees, in which the work of the individual for the community had become automatic action, he argued that the morality of working for the common good had slowly evolved as an instinct among humans" (Sapp 1994, 61–62).

The organicist paradigm became important not only for the nineteenth-century evolutionary theories but also for ecological theories of the early twentieth century. Within these ecological theories, biological analogies were also utilized, especially those with social insects as models. Both sociologists and entomologists used these analogies, as will be discussed in detail later. Donald Worster ties the end of the organicist paradigm to the specter of totalitarianism and the outcome of Hitler's Germany: "Confronted with the example of Nazism, many organicists began to retreat a few paces from the integrative ideal. That kind of state was not at all what they or nature had meant by 'relatedness.' Their organismic model was, or should have been, intended to be less centralized, less dominated by a single directive power" (1985, 330). Precisely because social insects had been the integrative example par excellence, ecologists needed to distance themselves from these particular analogies. Sociological theory moved away even more definitively from the use of biological analogies that were associated with organicism or social Darwinism. This may be, in part, a reaction to the program of eugenics during the Holocaust, but it was also associated with the move toward a functionalist and quantitative dominant perspective for the discipline. This separation became more and more significant for sociology during the end of the nineteenth century (Martindale 1960, 42). The ecological influence within the discipline staved off a complete separation until the 1940s. The entomologist William Morton Wheeler also noted the separation of sociology from the previously shared analogies but indicated that biology would still feel comfortable using such comparisons, as humans remained a part of their domain of study (1928a, 303–4). Comparisons between the social structures of insects and humans will be placed within this historical context of organicist, evolutionary, and ecological theories because this highlights the co-construction of models of social organization and sociality within both disciplines.

Sociohistorical Methods

Using sociohistorical methods presents a different "story" of sociology and entomology than would an internalist history of science approach. As stated earlier, I see both natural and social science as social constructions that incorporate political and cultural aspects. Presenting historical context requires more than relaying names and events; it involves delineating the complicated processes that have occurred. While the larger historical context of evolution, ecology, and organicism needs to be understood, my focus lies within the disciplines of sociology and entomology and the history that they share. Because distinct disciplinary boundaries did not exist until the 1940s, the interactions between biology and sociology are rooted in the intellectual history of both disciplines. Although the general interactions can be studied, the interactions between the specific disciplines of social science and natural science provide for a more detailed account (Cohen 1994; Mirowski 1994). As sociologists of science have indicated from Thomas Kuhn onward, science is rife with political and personal biases as well as negotiations (Knorr-

Cetina 1981; Latour and Woolgar 1979; Mulkay 1985; Richter 1972). Serious epistemological issues arise from any human interpretation of social insects. To begin to see how serious these issues might be, it is important to understand the history of the classification and terminology used in entomology today, as it is interwoven with the history of social theory.

Social Constructionism

Although there is ongoing discussion in both the natural sciences and the social sciences over the use of anthropomorphism, analogies, and classification systems, there appears to be a sense that these are as "natural" as the phenomena they describe. For instance, the classification schemes for social insects tend to be thought of as reflecting the objective reality of insect behavior. However, the process of deconstructing these classification schemes makes apparent the sociopolitical aspects of their creation. In order to make something seem more like a natural fact, its sociopolitical aspect becomes disconnected from the construction (Latour and Woolgar 1979). Both the social realm and the natural realm reinforce certain ideas through their continued embedded interaction. Terminology and analogies that have been used in both the discipline of sociology and entomology should be questioned rather than accepted as fixed. Tracing the social construction of these terms and analogies is a helpful methodological strategy for deconstructing the larger discourse of hierarchical classification systems.

Social constructionism has been accused of subscribing to an extreme relativism. Sandra Harding claims, however, that the social constructionism she uses in standpoint theory illustrates the relativism within *all* knowledge claims and actually calls for "stronger standards of objectivity" (1998, 18). Although standpoint theory recognizes the influence of social location in the creation of knowledge claims, it does not present itself as unable to critically assess those claims; if anything, critical assessment becomes more stringent because it is grounded in the evidence of social construction. I agree with Harding's assertion that acknowledging standpoint is not the same as taking a completely relativistic stance, and this informs my use of a social constructionist method.

Acknowledging that various standpoints contribute to the social construction of knowledge claims opens the door to voices of opposition to the dominant discourse. Understanding social location can enhance the analysis of varying interpretations within the discourse. Distinct patterns of this variation can be viewed as oppositional discourse that contrasts with the dominant discourse. The realization that scientific discourse is created by actors whose social locations fall within the intersection of race, class, and gender precludes endowing the dominant discourse with a deterministic, universal quality. Shapin makes a distinction between the "institutionalized cosmologies" and those that are subversive. "[D]iscontented social groupings may elaborate cosmologies which are not 'like' the dominant institutions in which they live, but which are, rather, signals of an ideal social order or tools crafted to subvert the dominant order" (1979, 47). Oppositional discourse contains naturalizing analogies as frequently as does the dominant discourse. And while oppositional discourse is not equal in power, it stills offer a challenge to the dominant discourse. Identifying and presenting this oppositional discourse alongside the dominant discourse serves as a reminder that the dominant discourse is neither monolithic nor the universal or natural expression of knowledge claims. Combining the evidence of the social constructionist approach with a sociohistorical context can considerably strengthen the use of critical discourse analysis.

Critical Discourse Analysis

New perspectives within science studies of classification systems, the cognitive sciences, and the field of ethnobiology dealing with classification are building support for the idea that classification and description of the natural world cannot be separated from social location and social institutions. This calls for a different type of analysis that consciously acknowledges such a connection. Critical discourse analysis (CDA) can aid in understanding how knowledge is presented or contested and how dominant ideologies are constructed and reinforce dominant institutions. This type of analysis recognizes that language should be interrogated for its role in constructing knowledge and practices. Biological analogies, classifications systems, specific concepts, and terms contain underlying assumptions and constitute ideology that gets bound up in institutional structures and practices (Fairclough 1989; Gilbert and Mulkay 1984; Lewontin 1991; Mulkay 1985; Thompson 1987). Questioning the generally accepted assumptions found in the discourse of sociological theory and entomological realms provides an opportunity to assess its effect on disciplinary fields and social structures. A critical discourse analysis addresses the meanings of these texts on several levels: as dominant paradigm for the entomological classification of social insects; as discourse that emerges from competing knowledge claims; as shared discourse between the social sciences and the natural sciences; as oppositional discourse; and, finally, as a discourse that is interwoven with the social structures it attempts to legitimate and draw legitimation from.

Unlike other types of discourse analysis, CDA does not abstract language to a level that is disconnected from social structure. Discourse is not seen as a cognitive and hardwired process. This method of discourse analysis is more in keeping with the claims that I am making that challenge the idea that classification systems are a simple reflection of natural properties. There is a match between the goals of critical theory, critical science studies, and critical discourse analysis because of a shared interest in epistemological clarity. Because CDA's goal is to illuminate the interconnectedness of things, there is a central assumption that discourse is connected to social structures or institutions. In this sense, language is seen as having to do with power and ideology, and the method consciously adopts this perspective (Fairclough 1995).

There is a dialectical relationship between discourse and institutions, and critical discourse analysis addresses this by taking into account the nature of that relationship. Hierarchical discourse using social insect analogies that were drawn from hierarchical social institutions in turn helped create and reinforce those institutions through naturalization. Hierarchical institutions have real consequences for those situated within them, whether those affected hold positions of privilege or oppression. Discourse can help to maintain these positions or challenge and sometimes alter them.

The potential for changing the social structure and discourse is assumed in a critical discourse analysis. This change can come from within the current dominant discourse or with help from the act of critique itself. Critical discourse analysis is a method involving the critique of text, but it is also involved in the praxis of change stemming from the denaturalization of social structures as depicted in the dominant discourse. The goals of CDA are to "unmask and delineate taken-for-granted, common-sense understandings, transforming them into potential objects for discussion and criticism and, thus, open to change" (Phillips and Jorgensen 2002, 178).

One change that critical discourse analysis could offer the dominant discourse of insect sociality would be to challenge the notion of hierarchy in the very definition of eusociality and in some relational terms such as "queen/worker" and "slave/slave-maker." In addition, the category of eusociality rests on assumptions that mirror ideas of a colonial worldview and modern organizational structure, with a specialized division of labor being the key to identifying certain types of insect organization as more complex. Newer conceptions of social organization challenge this emphasis on a specialized division of labor and instead focus on self-organizing properties. This is but one significant challenge that brings up a larger question: Should organizational structure and behavior be ranked on high/low standards of complexity, with one ranked highest and seen as a standard against which to measure others? The new self-organizing model is beginning to be used in analogies that encompass both natural and social organizations; changing the previous understanding of organizational models for social insects appears to have contemporary implications. Revealing the underlying assumptions behind the creation of past concepts and analogies can inform thinking about other new models that may be embraced in the future. If a new self-organizing model becomes accepted, what will this mean for the naturalization of social structures? As Sommer notes, changes in scientific discourse can help promote changes in the relations of society (2000, 20). Sommer's work deconstructs the scientific discourse involving primates as models for human organization. My research on social insects and scientific discourse has similar goals; it also has similar potential to change views of relations that are currently viewed as "natural."

A critical stance is imperative in order to inquire into a previously accepted scientific discourse concerning insect sociality. Critical theory and methods can provide a way to explore the legitimation loop that occurs between the shared analogies that compare social insects and humans. Understanding that theory, classification, and language are all intricately interwoven with actual social relations takes one beyond the assumption of an innocent anthropomorphism or the scientific prerogative of objectivity claims.

3

A Bee or Not a Bee
HISTORICAL AND CROSS-CULTURAL INTERPRETATIONS

Does the classification of social insects change depending on the social location of the ones defining it? I will illustrate that, indeed, classification and the use of analogies are socially constructed and influenced by the social location of those using these classification schemes and analogies. In making this claim, I do not intend to take a completely relativist stance; I believe that social insects organize outside of the interpretation of humans. However, how we know and describe this organization varies over time and place and hence is socially constructed. Humans may variously describe the organization of social insects as a rigid and hierarchical bureaucracy reflective of modern organization or, based on a more postmodern understanding, as self-organizing. However, the organization of social insects itself changes very little; it is our understanding of it that changes. These changes are not brought about simply by better methods of observation or scientific techniques, but more often entail a paradigm shift in how scientific knowledge is understood. This change in paradigm usually reflects a parallel shift in the understanding of human social organization and roles. Because of this connection between paradigm shifts, interpretation, and larger social context, scientific knowledge about the organization of social insects can be considered situated knowledge. As previously discussed, feminist and postcolonial science studies locate knowledge in the intersections of race, class, and gender without essentializing these positions.

In addition to race, class, and gender, there are other factors to consider regarding the social construction of knowledge. An individual's political stance may contribute to what knowledge is considered legitimate, and even geographical location can affect viewpoints. State boundaries and political economies may influence constructions of knowledge through national differences or styles of doing science (Harwood 1987; Hess 1995). Postcolonial science studies combine many of these factors to describe how the dynamics of colonization and ongoing development policies alter scientific understandings. Indigenous science has been labeled as folk knowledge or incorporated without credit into the canon of Western scientific knowledge; in response, postcolonial science studies have begun the process of reclaiming and also critiquing Western science. This contribution has also emphasized varying standpoints and the resulting cross-cultural interpretations that challenge a universal Western model.

Sandra Harding proposes a feminist, postcolonial stance from which to explore the social construction of science, claiming that it has and continues to be a multicultural social endeavor (1998). I also take this stance concerning the social construction of the classification of social insects. This social construction can be considered a co-construction alongside human models of social organization because of the shared analogies that have been used to bolster both interpretations. Therefore, examining the scientific classification of the organization of social insects has import for the various ways in which human social organization has been socially constructed as well, especially as naturalized by comparison. Although the co-construction is a mutual process, we can argue only that human social institutions are naturalized and legitimated by their comparisons to social insects, and that our scientific understanding of social insect organization is affected by the co-construction, but not that the actual organization of insects themselves is altered. In fact, contextualizing this becomes the greatest argument of all for social construction.

UNNATURAL THEORY

The connection between the various constructions of social insect organization and sociology lies in the relationship to theory. Social theorists used biological examples in their theories of human social structure; from the organicists of the early nineteenth century to the sociobiologists of the twentieth and twenty-first centuries, insect societies have been considered to offer a valid, evolutionary comparison to human societies.

These theorists and others were (and still are) a part of the process of creating a sociopolitical classification system for social insects as well as an overly deterministic understanding of organizational structure for insects and humans. The varied interpretation and use of biological analogies in theory points to their social construction and the need to explicate the deeper epistemological issues involved. As Escobar (1992, 62) notes: "The neutral, 'positive' character of theory can no longer be taken for granted. The questions of who speaks and from what institutional sites are only the starting points of the inquiry." Understanding where a particular theory "comes from" is a step toward better understanding its legitimating force and its sociopolitical implications.

Social theorists such as Herbert Spencer, Peter Kropotkin, Charlotte Perkins Gilman, and Franklin Giddings all used examples of insect societies to stress theories and concepts of social evolution and social structure. However, whether these examples stressed competition or cooperation as the key to social evolution varied as to the theorist and their political orientation. For example, in step with many Russian scientists who did not accept a "Malthusian" conception of Darwin's evolutionary model (Graham 1993; Mikulinsky 1981; Todes 1989; Young 1985), Kropotkin presented an interpretation of insect social structure based on cooperation, as evidenced in his 1902 book *Mutual Aid*. This interpretation was influenced not only by his sociopolitical location, but also by his experience with the natural world seen through the geography of Russia, as Kropotkin was a state geographer for a time. The conditions for survival that he noted seemed to require group cooperation rather than competition (Gould 1997; Kropotkin 1902). In contrast, Herbert Spencer, who embraced Malthus's ideas of population and who, like Malthus and Darwin, claimed England as his homeland, believed that insect social structure reflected the "survival of the fittest," a term he coined. His ideas about insect evolution paralleled his views on social evolution as linear development from simple to complex organizational forms. This evolutionary scheme was praised by Alfred Russel Wallace at the time and is still influential for studies on insect evolution today.

It is clear that theories of nature are sociopolitical sites of knowledge. They may reflect the theorists' overt political agendas or the sociopolitical environment within which they happen to be located. Engels claimed that Darwin's theory of evolution was a direct grafting of political economy and competition onto nature, which was then used as a means to justify a competitive human social structure. Doubting the neutrality of Malthusian-inspired evolutionary theory, Engels accused Darwinism of naturalizing a political stance that favored bourgeoisie, laissez-faire capitalism. And while Engels understood the power of such a legitimating loop, he argued that this political stance could not (and should not) be conceived of as natural law (Engels [1883] 1940, 186–87).

Political views are usually not the unique product of individuals but rather express attitudes reflective of the larger context of time and place and tend to be contested by various groups. Likewise, as Kuhn described, scientific theories or concepts represent dominant paradigms that may be challenged by others. In the typical historical representation of scientific knowledge, these disputes may be glossed over and scientific knowledge can be made to seem like one unbroken discovery of universal truth revealed by individuals. Thomas Kuhn's classic work on paradigm shifts pointed to a more subjective and sociopolitical process involved in the advance of scientific knowledge. Critical feminist and postcolonial science studies identify multiple standpoints that contentiously create scientific theories. The intersections that create these standpoints are complex and dynamic and yet can be analyzed as positions that generate particular sets of knowledge. Scientific theories are a distinct product of time, place, and personal biography, and alternate theories can be formed simultaneously by those who are situated in different social locations.

There is strong evidence of alternative vantage points for the classification of insect social structure, especially as utilized in analogies for human social structure, because not all accounts mirror a Western view of the organization of social insects. Vantage points are difficult to categorize, and the examples I use are loosely grouped into those that are historical, gender-related, and cross-cultural. These areas contain intersections with class, race, sociopolitical factors, national boundaries, and geography.

HISTORICAL SOCIAL CONSTRUCTION OF SOCIAL INSECTS

As noted in chapter 1, the definition for a social insect has developed over time, becoming more specific and adhering closely to a Western bureaucratic style of organization. Other changes over time illustrate that biological classification and terminology involving social insects is subject to social construction. For instance, early entomologists identified the "queen bee" as a "king." Although in 1586, Luis Mendez de Torres of Spain wrote of the possibility of a queen bee, as did Charles Butler, who wrote the *Feminine Monarchie* in 1609, it was not until Jan Swammerdam's work in 1740 that the identification of the head of the colony became scientifically confirmed as a female "queen" bee (Burke 1997; Crane 1975; Ransome 1937; Root [1891] 1966). Previously the interactions of the colony were used to illustrate how people should obey the king (Ransome 1937). Even after the analogy shifted to a queen and her followers in a monarchy, political obedience to a king was still implied (Merrick 1988). Burke (1997) argues that it was not only the new technology of the microscope but also Swammerdam's lack of investment in the political structure of the monarchy that allowed him to see past the assumption of the bee as king, thereby realizing that the sex of the prominent bee was female.

While most European entomologists seemed to embrace the imagery of a female queen, U.S. entomologists, whose work became influential in the late nineteenth century, had difficulty with the idea of the "queen-as-ruler" analogy, or at times even the idea that females may have more status than males at all. Therefore the intersection of history, politics, and national boundaries had implication for another historic shift in the social construction of the queen's role. Analogies involving social insects in America came to embody the cultural values of the newly emerging nation in the eighteenth century (Barnes 1985; Withington 1988). The U.S. entomologist Edwin A. Curley expressed discomfort at having to describe female status in this society of insects in his 1885 *Entomologica Americana* article, "Bees and Other Hoarding Insects: Their Specialization into Females, Males and Workers," writing: "I beg that you will note the order in which I have mentioned the sexes. It is females and males and not males and females. In a scientific discussion, I feel constrained to tell the unvarnished truth regardless of the consequences to the social fabric; and among the Hymenoptera, it is most certainly a fact, that the ladies are all-important, and the gentlemen approach as closely to perfect insignificance as it is fairly possible to conceive" (1885, 61).

Thus an impulse toward scientific accuracy compelled Curley to explain the division of labor as emphasizing female insects over males, despite his misgivings about "the consequences to the social fabric" that such an analysis might entail. Since, as previously stated, the "queen as ruler" analogy appeared to be problematic for U.S. entomologists, Curley's social location as a U.S. entomologist may be an important aspect of his interpretation. Some of the problems with this concept were resolved in the way illustrated by Vernon Kellogg's reframing of the queen's status: "The queen is no ruler; she is the mother, or, better, simply the egg-layer for the whole community" (1908, 526). The description of the queen as not even quite a mother but more "just an egg-layer" focuses on her "function," but diminishes her status.

The intersection of history and gender can be viewed in the concern expressed by women scientists with the queen's new role as "egg-layer." Charlotte Perkins Gilman, a U.S. feminist and sociologist writing in the late nineteenth century, used the analogy of the reproductive division of labor among insects to illustrate women's economic dependence on men. This issue was specific to her social location as a white, middle-class woman from a time period that fostered dependence for women such as herself. She predicted that, "In any animal species, free from any other condition, such a relation would have inevitably developed sex to an inordinate degree as may be readily seen in the comparatively similar case of those insects where the female, losing economic activity and modified entirely to sex, becomes a mere egg-sac" (1898, 58). She believed that precisely because there was a socially constructed and rigid division of labor between the sexes, a socially unhealthy dependence was created. Citing the social insects as an extreme example, she illustrated her concern that this state may be occurring to human females because of the exaggerated and gendered human division of labor that she universalized. She did note, however, that in the case of social insects, that dependence was not on males but actually on other female workers. "The female bee and ant are economically dependent, but not on the male. The workers are females, too, specialized to economic functions solely" (1898, 6).

Does this dynamic of female queen and female workers make a difference? According to Anna B. Comstock, a female U.S. entomologist, it did. Comstock noted the division of labor in what she termed a "socialistic" community of insects and decided that this division of labor was necessary under socialism so that females can be workers. "[I]n the socialistic bee-community the work is carried on by unsexed females. It evidently has not been a part of the true economy of the perfect socialism to unite motherhood and business life in one individual; therefore, a division of labour takes place. The queen mother is developed into a highly efficient egg-laying machine, while all her worker sisters remain undeveloped sexually, and thus have time and energy to devote themselves to bringing up the young, keeping the house, getting the food, and administering the affairs of the body politic" (Comstock 1905, 39–40). The mother is able to be "an efficient egg-laying machine" while the "worker sisters" have "time and energy" for their tasks. There is a sense that this works out a bit better for the worker sisters but that it is still a cooperative venture amenable to both, which is made evident by Comstock's use of the term "socialistic" (her meaning, described elsewhere in her text, is of a utopian socialism).

Comstock also noted the stigma attached in human society to a woman who works without being a mother, whereas in the bee community, a worker is not expected to be a mother. She found that, "it is interesting to note the differences in prejudices that obtain in the hive and in human society. In the latter we regard it as scandalous when the female, avoiding the duties of motherhood, goes abroad gathering honey and pollen at her own sweet will; but in bee society it is not merely a scandal, but a misfortune, when the worker bee has ambitions to be a mother" (1905, 45). Although not considering herself to be a feminist, Comstock noted this "prejudice" in human society that accompanies a woman who chooses to not be a mother but rather to be "gathering honey and pollen at her own sweet will." She observed that this is considered scandalous in human society as opposed to insect society. Comstock herself may have felt stigmatized for this choice as she was childless and chose a life of work in entomology alongside her husband at a point in history when this was considered unconventional behavior. The work of both Gilman and Comstock reflects particular historical time periods as they intersect with gender roles of that time. Their gendered social concerns are found expressed in analogies or interpretations of social insect organization.

There seem to be ongoing changes in the role attributed to the queen in insect society. In more current interpretations of insect colonies as self-organizing, the queen's role diminishes even more as her connection to the actions of workers is seen as nonexistent. And yet, despite this recent threat to the queen's title, other emerging views foretell of a return to the queen's elevated status. In the *Genome News Network* in 2003, Kate Dalke writes of "the queen bee's allure," claiming: "The queen honeybee is irresistible to worker bees. Now we know why. The queen, who produces the most complex pheromone known in the animal world, has at her command a cache of chemicals to make other bees do her bidding. She uses this powerful and seductive perfume to attract a retinue of worker bees that lick and groom her and carry chemical signals back to the rest of the hive." It seems as if we might as well proclaim, "Long live the anthropomorphized queen!" so embedded are the qualities associated with the term "queen."

THE INFLUENCE OF GENDER

During a time when male scientists and social theorists used examples of social insects to illustrate the gendered division of labor, Lydia E. Becker, the nineteenth-century suffragist and advocate for science education for women and girls, presented a controversial essay that depicted social insects a bit differently from the norm. "Is There Any Specific Distinction between Male and Female Intellect?" was read at the Manchester Ladies Literary Society (a women's science society) as well as the British Association for the Advancement of Science and was published in *Englishwoman's Review* in 1868 (Parker 2001; Rowold 1996). The piece began by noting that some authors had used biological evidence of differences in physical strength as an indicator of intellectual differences between men and women. Becker then proceeded to offer evidence to the contrary. Revealing and emphasizing the role reversal found in the structure of a bee colony, she stated:

> In the hive bee, the ingenuity to work and the power to govern are vested in actual or potential females, while the males, kept within their "sphere," are ignominiously hustled out of existence, whenever they are tempted to step beyond it, or unreasonable enough to ask for a share of the good things of the hive. But it seems fair to conclude that this subordination is not due to any inherent inferiority in masculinity, as such, but simply to the fact that their bodily organisation leaves them defenceless against the terrible weapons of the superior sex. Had nature gifted the

male with stings, they would assuredly assert their right to live on equal terms with other members of the community. (Becker 1868 in Rowold 1996, 17)

Despite that the inactivity or the insignificance of the "male drone" was often described in the literature, Becker's separate sphere analogy is unique and expresses her feminist standpoint in attempting to turn the typical naturalizing of role inequality on its head.

If there were differences in the status of the queen for European and U.S. entomologists, on some matters concerning the reproductive division of labor they could more easily agree. The marriage flight of the queen took center stage in the descriptions of the reproductive division of labor as the queen established herself at that point, and except in the case of termites, otherwise known as white ants, the male drone dies after mating. It might be presumed that this would be depicted as the most important moment in *his* very short life, but the literature tends to describe the event as the female queen's "shining moment" and with other such phrases. The following description of termites by the U.S. entomologist Frank E. Lutz includes a judgment of the female insect's attractiveness as important for her success finding a mate during her marriage flight. This description is then grafted onto human females, along with a word of advice to follow the ways of the termite in these matters.

> Males and many virgin females fly out together. There is more or less courtship during the flight but no actual mating then. Soon the females settle to the ground and shed their wings, the flight being over. If a female has been at all attractive, at least one male is with her and, after the wings have been shed, courtship becomes more energetic. Matters are finally arranged and the happy pair starts housekeeping, the male living to enjoy home-life. Female "sluggards," if such there be, among humans may consider the ways of real ants and be wise, but most men would doubtless prefer the ways of the so-called white ants, the blondes. (Lutz 1941, 114)

W. P. Pycraft, a European entomologist, describes a very timid and demure young queen whose suitors seem to be in awe of her during what seems to be described as only *her* nuptial flight: "the young queen ventures abroad, timidly at first, to stretch her wings in the sunshine. She is preparing for the great moment of her life, the nuptial flight. So far, though drones may swarm on every side of her, no sign of recognition is given, nor do the males evince any consciousness of her presence. She behaves warily and demurely throughout" (1913, 278). The males fly with her, but the event is evidently the significant marker for her life, not theirs.

These descriptions of the queen and the marriage or nuptial flight are numerous in the literature and are typical of the dominant discourse. As Emily Martin has similarly suggested about depictions of the female egg and male sperm in the human reproductive process, these scientific descriptions mirror stereotypical gender roles. Martin found that, even when new scientific information became available that defied these stereotypes, the only change in the depictions of the role of the egg was that it was portrayed as seductive rather than as passive (1991). Similarly, descriptions of the queen and the reproductive division of labor are laden with stereotypes in the entomological literature and the sociological literature that uses entomological analogies. The reproductive division of labor is predominant in the literature as this is seen as one of the key characteristics of eusociality. From that foundational emphasis, the "marriage or nuptial flight" becomes the first step toward the new queen's life that will be busy fulfilling these attributed stereotypes. This transition is often featured within entire book chapters, and the descriptions are wrought with gender stereotypes.

Charlotte Perkins Gilman, in a foreshadowing of Martin's concerns, addressed the unbalanced focus found in both scientific and nonscientific literature in the nineteenth century. In explaining the male author's fascination with human sexuality, she specifically used an analogy of bees, parodying the female standpoint (bees) as well as the males (drones).

> If the beehive produced literature, the bee's fiction would be rich and broad, full of the complex tasks of comb-building and filling, the care and feeding of the young, the guardian-service of the queen; and far beyond that it would spread to the blue glory of the summer sky, the fresh winds, the endless beauty and sweet-

ness of a thousand flowers. It would treat of the vast fecundity of motherhood, the educative and selective processes of the group-mothers, and the passion of loyalty, of social service, which holds the hive together.

But if the drones wrote fiction, it would have no subject matter save the feasting, of many; and the nuptial flight, of one. To the male, as such, this mating instinct is frankly the major interest of life; even the belligerent instincts are second to it. To the male as such, it is for all its intensity, but a passing interest. In nature's economy, his is but a temporary devotion, hers the slow processes of life's fulfillment. (1911, 99)

Gilman is stressing that human sexual love or the insect marriage flight, as the case may be, is overemphasized, and she believed that other observations and perspectives were left out due to the focus on this one aspect of life. Her work did not contain romanticized versions of the marriage flight and tended toward emphasizing the communal aspects, even going so far as to coin the term "group-mothers." This description of the social structure of bees also parallels her suggestions for changing human social structure found in her book *The Man-Made World; or, Our Androcentric Culture*.

This small sample of quotes from sociologists and entomologists illustrates that social location influences how gender and the reproductive division of labor may be portrayed. The dominant discourse reinforced traditional ideas of the reproductive division of labor, with very rigid gender roles as evidenced in both the early and later entomological descriptions. Oppositional discourse attempted to take the same behavior and interpret it in a way that challenged gender inequalities. That being said, it should be noted that it was mainly women who were nature writers in the nineteenth and early twentieth century, and most borrowed information from science that generally reinforced stereotypes. As standpoint theory suggests, oppositional discourse would come from an achieved perspective, not simply a happenstance of social location that comes from birth. Dominant or oppositional discourse can come from various quarters that emerge from an intersection of race, class, gender, and other external factors.

Radhakamal Mukerjee is an example of a theorist approaching social insect analogies from a non-Western male vantage point. Mukerjee was an Indian sociologist who was theoretically situated in the field of human ecology. Mukerjee's non-Western standpoint produced interpretations of entomological evidence that differed from those usually found in the dominant discourse. His ideas about social insect queens contained references to exploitation and colonization: "Some species are entirely dependent upon their slaves in whose nests their young queens establish their own brood. The character of this exploitation thus varies. . . . The swarming activities of bees, wasps and ants resemble the marauding and colonising expeditions of men while the development of highly specialised social parasites with no workers which live at the expense of other species resemble human societies which subsist by maintaining a slave or proletariat population to a necessary level" (Mukerjee 1942, 34).

According to Mukerjee's account, the patterns of exploitation are similar for queens who use slaves and for human societies who use slaves or a proletariat population. Although the literature in both sociology and entomology described these situations of queens with worker-slaves, there was generally little criticism of this arrangement. Mukerjee also described the mating ritual or marriage flights of social insects not in romantic terms but rather as a colonizing expedition, which is significant since he was writing from the social location of a sociologist living in country that had been colonized. His remarks stand out in the literature because critical stances such as these were not often expressed in the analogies using social insects or in the descriptions of their organization and behavior. In contrast to the U.S. entomologist Frank Lutz, who believed that the gender roles of termites might offer useful instruction for human females, Mukerjee found that the model might be more suited to changing human male behavior:

> It is only in the termite and human societies that we find males consorting with females in the colony throughout life and that workers and soldiers are of both sexes. The male has been socialised in the termitary but not in the human society. It is probable that man's social evolution in the future will chiefly consist of the chastening and discipline of sex and aggressiveness on the one hand, and the utilisation of power and vitality of the human male for constructive purposes. Man today is too much of a destroyer,

a law-breaker and war-maker to assure the safety and continuity of his species. Can he reach the social subordination of the male as in the termitary, the most perfect type of society that has evolved in nature, without sacrificing his plasticity? (1940, 48)

The intersection of gender, historical time period, and culture can alter the description and classification of social insects. The use of social insects as a model can vary according to the interpretation put forth, and this interpretation is influenced by many social factors. The methods utilized to observe social insects or to describe their behavior may also vary and bring about terms and relationships that conflict with the dominant Western classification.

This is exemplified by Maria Sybilla Merian's approach to the study of insect life in the seventeenth and eighteenth centuries, work that was especially remarkable for a woman of this time period. Already established for her work in entomological and botanical illustrations, Merian traveled to Surinam, a Dutch colony, in 1699 along with her daughters to collect insect and plant specimens for her book *Metamorphosis Insectorum Surinamensium*. Margaret Alic (1986, 109) claims: "Maria Sybille [sic] Merian was one of the earliest entomologists. She was also one of the finest botanical artists of the period and a founder of biological classifications." A more detailed look at Merian shows how her work methods actually deviated from traditional Western classification.

Merian was most renowned for her illustrations of interactions between insects and their host plants, a unique method of observing and depicting insects in their environment. She also diverged from European descriptions by including indigenous terms and descriptions of insect behavior, most often gleaned from the local women with whom she had established relationships. Another outcome of these interactions was that Merian noted that some of the indigenous women used local herbs to abort children rather than have them born into slavery; she in turn criticized the Dutch colonists for their slavery practices. As Davis (1995, 187) remarks of Merian's unique perspective: "Just as she did not classify the species of flora and fauna, so she did not classify the customs of Amerindians and Africans." For the time period, refraining from both was unusual. One term that she used that differed from the European term and description of behavior was "visitation ants," used to describe the particular group of social insects still known today in Western classification as "army ants." Merian was using the indigenous expression that depicted the behavior of the ants as "visiting" houses to systematically rid them of pests. This interpretation was based on the interaction that indigenous people had with the social insect, whereas the later term, "army ants," was developed by the non-native colonizers, who interpreted the behavior of this species of ant as mimicking the actions of colonizing armies.

CROSS-CULTURAL INTERPRETATIONS

Differences in cross-cultural classification of social insects can serve as evidence that the dominant classification scheme is arbitrary and socially created. Several studies of indigenous classification systems suggest that the social organization of insects is viewed differently from the Western conception that has become standardized in modern science (Bentley and Rodríguez 2001; Ellen 1993; Gurung 2003; Novellino 2000; Posey 2002; Wyman and Bailey 1964). Some ethnobiologists note that cross-cultural comparisons can be difficult because of a lack of shared cultural assumptions. However, what is obvious from the literature is that there are different interpretations of the organizational structures of social insects and various types of comparisons to human society, which is enough to show social construction.

It would be a mistake to think of indigenous knowledge or classification schemes as static or unadulterated by any outside influence. The examples I use are not intended to suggest some natural interpretation uncluttered by "civilizing" forces; rather they point to the fact that there are some cultures that do not share Western assumptions. The indigenous classification is not any more "pure," objective, or correct; it is just a different vantage point of interpretation, one that reflects the various standpoints within that culture at a particular time period. Ethnobiologists often attribute a cultural context to indigenous classification as if this were unique to indigenous society. As Posey states, "Folk taxonomies are in and of themselves cultural statements, but it appears that these taxonomies may reflect deeper cultural patterns" (2002, 92). Postcolonial theory would argue that calling indigenous systems of classification "folk taxonomies" marginalizes them and contrasts them to Western scientific classification.

The initial emphasis of postcolonial science studies was to attempt to emphasize the scientific nature of indigenous knowledge and elevate it to the more objective status of Western science. Recently more emphasis has been on critiques of Western science or ways of incorporating transnational understandings (Ancarani 1995; Nandy 1988; Prakash 1994, 1999; Reid and Traweek 2000). Another possible approach is to show that Western scientific classification can also fit the definitions ascribed to indigenous knowledge systems. For instance, Posey describes the conclusions one can make from studying indigenous classification with the following statement: "It appears that belief systems can play an important role in classification patterns and that such patterns can, in turn, offer an emic guide to cultural realities of perception" (2002, 92). It would be equally accurate to apply this statement to Western classification. All classification should be subject to an analysis of the cultural factors that influence it. This can be done through revealing the social construction of classification as a process.

Analyzing the connection between natural models and social structures should not be restricted to the study of indigenous cultures. The questions that Posey claims should not be avoided in regard to Kayapó culture echo some of the questions that science studies ask of Western culture as well. To what extent does a perceived natural model influence the creation of social structure? Or, rather, is it the social structure that influences the perception of the natural model? The answer to these questions for both indigenous and Western cultures undoubtedly involves co-construction. This co-construction will not be universal, but rather the result of the interplay of culture and the perceived natural model. Many factors will influence the co-construction, including what is available within a given environment from which to derive a model or how the culture is defined and by whom.

Wyman and Bailey's work on the Navajo system for classifying insects was written in 1964 and therefore reflects the classification of that time. It cannot address any changes that may have occurred in contemporary classification, and, for our purposes, it need not do so. All classification changes in any living culture, so any classification system is always located in a particular time period. What is instructive in the following descriptions is the variation in interpretation, illustrating that classification of social insects is a socially constructed activity.

Historical and Cross-Cultural Interpretations

There is evidence that Navajo analogies between the social organizations of insects and humans reflect the Navajo culture as uniquely as Western interpretations reflect a distinctly Western culture. Wyman and Bailey report that, according to Newcomb (1940), the Navajo claimed to have modeled their housing design on that of the "ant people" in their legends who were the first home builders (1964, 46). Ant people and up to a dozen or so other insect people figure in the creation story or emergence myth of the Navajo; these insect people populated the "first dark underworld." The Red Antway is a chant complex that describes the emergence of the ant people from this dark underworld to the upper world. Bees and wasps are featured in other creation stories (Wyman and Bailey 1964).

In the study of the Navajo system of insect classification, researchers presented their informants with dead insects in boxes. However, because the concept of movement is very important for the Navajo classification of insects, the informants' ability to identify the insects was hindered by the researchers' use of dead insects (Wyman and Bailey 1964, 23). This distinction between dead insects and live insects identified in their habitat is no small matter and is not just an indigenous issue. In struggles over methods for Western taxonomic methods, William Steel Creighton challenged William Morton Wheeler's classification schema of ants in regard to this same issue. Creighton claimed that the dead insect collection used by Wheeler did not offer the same information as what was found with fieldwork (Blu Buhs 2000). In the Navajo study, informants also tended to ask the researcher where the specimen had been found, which indicates that place is important for identification as well. Insects were matched with their behavior and habitats whenever observations were presented to researchers (Wyman and Bailey 1964, 26).

The insect's perceived behavior is used in Navajo descriptions. Western classification also uses perceived behavior to give terms to social insects; as mentioned earlier, ants are termed "army ants" because their behavior is perceived to be similar to the movement of an army. Other ants are termed "slave-makers" or "slaves" based on a dynamic that is perceived to be similar to human slavery. However, because the Navajo do not perceive ants to engage in either of the above set of behaviors, the Navajo classification of social insects does not include these terms. Alternately, for the Navajo there are perceived behaviors for ants that do not appear in Western classification; for instance, an ant that is perceived to be listening is termed a "listening ant." Although size, description of behavior, and movement is important in Navajo classification, color is not as important. After giving the term for "black ant," the researchers clarify: "Navahos [sic] are using 'black-ant' or 'red-ant' to mean simply 'ant' in general" (Wyman and Bailey 1964, 45). They also do not make a distinction between the sex of insects, according to the researchers: "Their treatment of insects . . . differs in a number of ways from the classification of plants. . . . One is that in naming insects they seldom introduce the sex duality without prompting and, indeed, when questioned admit they know little about it" (Wyman and Bailey 1964, 17). Without significant markers of color or sex it would be difficult for Navajo descriptions of social insects to include the type of analogies, behavioral observations, and terms used in Western classification. Social organizational analogies that include a queen, king, male drones, and nonreproductive females would not be relevant for Navajo models. The Western terms "slave" and "slave-maker" are not used because they do not match perceived behavior. But also it would be difficult to transfer this Western concept because demarcation between slave and slave-master also relies on color differences, with mainly red ants or "blond" ants enslaving black ants.

Without distinctions of gender or color (which stand in for "race" in some Western descriptions), it makes sense that gendered and racialized terms would not be used by the Navajo to describe social insects. This undermines the supposed naturalness of these terms and the analogies that are drawn from them; clearly they are a product of Western observers' cultural background. Wyman and Bailey credit another researcher, Gladys A. Reichard, with noticing the incommensurability between Navajo and Western categories. Navajo categories, in her observations, feature "duality, inclusiveness, overlapping" characteristics. Western assumptions about categories would need to be questioned, according to Reichard, were there to be any kind of comparison (Wyman and Bailey 1964, 16).

There is a lack of hierarchical ranking within some indigenous classification schemes that would make comparison to Western classification difficult. One example is provided by Ellen (1993), who suggests that with the Nuaulu of central Seram, it is not even possible to conceive of "deep hierarchies" in the classification schemes of animal social organization.

Historical and Cross-Cultural Interpretations 57

"The imagery of hierarchy, implying higher and lower, top and bottom, superior and inferior, with its technical associations of literate graphic representations and its social connotations of class and status, is absent from Nuaulu discourse on animal relationships" (1993, 89–90). The assumption of hierarchy in Western schemes of classification does not help explain the Nuaulu system or offer a foundation for comparison, Ellen maintains (1993, 89–90). This argument is similar to Reichard's earlier claim about the Navajo system.

When hierarchy is present in any classification system for social insects, it reflects the hierarchy within that particular human culture. An example of this is Darrel Posey's work on the social insect classification of the Kayapó Indians, who live in the states of Pará and Mato Grosso in the Amazon of Brazil. The Kayapó distinguish social insects from other insects, but they make distinctions among the social insects as well. The term "ñy" includes termites, ants, wasps, and bees, although they each have separate names. This overarching term separates them from other animals because of their social organization. Posey explains: "The ñy or social insects are seen to be in a special relationship to man because of their communal nature. All ñy colonies [villages] are thought to have a chief . . . and be organized into family units just like the Kayapó. They are known to have warriors and the sounds of the movements are likened to Kayapó movements and singing" (2002, 89).

The two most prominent social insects for the Kayapó are wasps and ants, as their behaviors are a model for certain roles within human social organization. While ants can be emulated for their hunting techniques, wasps are warrior role models (Posey 2002, 92). Wasps are especially important and their societies serve as a general model of social organization for the Kayapó. This is reflected in a creation story that centers on the Kayapó being a weak people living in the sky until they were able to study the ways of the wasp to gain knowledge of social organization and fighting skills. In addition, the nest of wasps is important because the Kayapó see it as representing the entire universe. "The hive is divided into parallel 'plates' that seem to float just like the layers of the universe. The Kayapó say that today they live on one of the middle plates. But in ancient days, they believe they lived on another plate above the sky" (Posey 2002, 90).

The social structure is seen as reflected in the nest of a wasp as it contains a sky layer and an earth layer within a circular universe. There is also a lower layer on which live all those deemed worthless. Termites live on this layer; they are seen as "worthless" because they are "weak" and "cowardly," unlike wasps and ants, and therefore they are not classified in as much detail as other social insects (Posey 2002, 92). The idea of layers structuring the universe is similar to the social insect model adopted by the indigenous Pälawan of the Philippines as described by Novellino: "Pälawan envisage a kind of cyclical system in which the seasonal production of honey depends upon the flow of bees from the upper levels to the central layer (tängaq tängaq) of the universe.... These animals are believed to be creatures of the over-world visiting the central layer of the universe 'to get the pollen from flowers.'... The flow of bees from the upper world to the middle level is said to depend on a number of conditions, such as a favorable negotiation between people and the Master of the flowers" (2000, 194).

Kayapó culture contains a hierarchical division of labor, and the detailed descriptions of the social organization of bees reflect this as well. Although the hierarchical division of labor is similar to Western classification, the terms for insect roles may vary. This is an interesting case for the social construction of terminology. In creating similarities between their social organization and bees, the Kayapó describe various positions in the hive, such as: "warrior bees"; "scout bees"; "worker bees"; "principal chief"; "subchiefs"; "wives of chiefs"; "young children of the chief." The terms and corresponding roles are as follows: warrior bees defend the nest; scout bees search for food and new locations, as well as help guard the nest; worker bees gather pollen; principal chief (some species are seen to have more than one) gives the orders for all activity; subchief relays the orders of the principal chief; wives of the chiefs are responsible for egg laying and child care; young children of the chiefs are taken care of and fed special food. Furthermore, each category of bees is seen to have its own subchief, and Posey believes these to correspond to the Western nonfertile queens. He also compares the principal chief to the queen and the wives of the chiefs to the Western "nurse bees" (2002, 119–20).

The interesting aspect of Posey's attempt to directly compare this Kayapó hierarchy to Western terms and structure is that Posey is comparing what he says is a male principal chief to a female queen. Perhaps the assumption that the principal chief is male is incorrect as the Kayapó have had female chiefs within their own social structure. Posey (2002) discusses the changes in gender status elsewhere in his study without noting how this may influence changes in classification. He claims that the Fundação Nacional do Indio (FUNAI), the official state institution that covers Indian affairs, "has favoured male leaders as the spokespersons for the villages. Consequently, female chiefs have disappeared and those males that speak Portuguese have climbed to positions of greatest importance" (27). Other changes in gender roles appear to be the desire to rebuild a "men's house" in the center of the village while women will live on the outside ring, which socially and physically structures an unequal involvement in the village. Despite the changing gender roles and the lowered status of females, the question still remains: could the "principal chiefs" of social insects be either male or female since female chiefs were a part of Kayapó culture? What does this mean in terms of gender and the "wives of the chiefs?" Has Posey's attempt to match these terms to corresponding ones in the Western classification blurred some of the categories and confused the original meaning of the indigenous classification?

All of the indigenous cases presented thus far may be seen to illustrate that terms and analogies differ according to cultural beliefs and social structure. There is no one perceived natural structure for social insects just as there is no one social structure for humans. Descriptions of social insects will be influenced by this variation in human social structure because culture and social location play a role in classification. Various cultures may place different levels of importance on certain social insects as in the example of the Kayapó emulating the behavior of wasps and disregarding termites as worthless. If the culture itself has hierarchical terms for their social organization, then those specific terms might be used to describe social insects. If the culture does not have any terms for hierarchical relationships, then social insects will not be seen to have hierarchical relationships within their social structure.

Are there any cross-cultural commonalities in the classification of social insects? Berlin (1992) claims that under a universalist understanding of classification, social insects (and certain other species) will always be named in every culture. Bentley and Rodríguez (2001) confirm these findings in their case study of rural Honduran ethnoentomology, although their study actually reveals that the reasons for naming may vary. What Bentley and Rodríguez cite as the "culturally important and easily observed" category into which social insects fall is instructive because of what actually makes social insects culturally important. For rural Hondurans, social insects are not viewed as important because of their "socialness," and nothing about their classification would indicate that this is even observed. The "social" insects are deemed culturally important for humans based on interactions that may be assessed as either helpful or harmful. Bees that produce honey are important as food, but also some are known as causing pain from stings. Ants are also a source of potential pain from stings or bites.

Perhaps because the group studied by Bentley and Rodríguez was comprised of farmers, their interest in insects was less abstract. Some groups within societies may be seen to have a particular relationship to insects, and farmers would be one group that might be thought to share similar views on insects cross-culturally. However, as Gurung (2003, 351) notes in a case study of Tharu farmers in Nepal, social insects are not even considered important enough by this group to classify with much distinction at all. The term "kiaraa" for insects includes a larger group of arthropods and any other creature that might be considered to cause "harm to crops, livestock or people." Ants are known only for their ability to pinch people, while bees are described as useful for pollination only if researchers press for a beneficial aspect of insects. Social insects are not used in analogies to human society, and insects in general are mostly viewed as "harmful or a mistake in God's creation." Insects that are, according to Western classification, subsocial or solitary insects feature more prominently in the Nepalese rural culture: fireflies, cicadas, and dung beetles appear in morality stories, while the rice bug and praying mantis rise to the level of spiritual rituals because of the intensity of their perceived harmfulness (365). Therefore, in his study of Nepalese farmers, the author disagrees with Berlin (1992) and focuses on naming based on the functional, utilitarian roles of the insects. He also questions Posey's (1979) methods for obtaining classification information that supports Berlin's assumptions (357).

Gurung (2003, 359) took note of gender differences in the way Tharu farmers classified social insects. Women tended to be more specific, naming more species, and described insects with the plant host. Gurung attributes these differences in part to the gendered division of labor. Both sexes had opportunities for observing insects, although through the lens of different tasks. The assumptions of general cross-cultural comparisons would attribute more homogeneity to the culture studied than may be accurate. As in every culture, there is a multiplicity of vantage points from various social locations within the culture itself.

Although this discussion of indigenous classification appears to simply point out that there are cross-cultural differences in classification, there are other issues at stake as well. One of these issues is that the researchers gathering information on indigenous classification work within the paradigm of the Western classification system. How does this influence the translation of information? Sometimes the entire research agenda is to compare identification with the Western Linnaean system of classification. When an informant doesn't specify a classification, sometimes even indicating that it isn't important, then researchers are often forced to guess or choose a category. This can result in "mistakes" in classification, particularly if indigenous groups do not place a species in a particular category that aligns to the Western one. There sometimes are elaborate explanations of myths and worldviews in order to situate the "mistakes" as having some alternative rationale. It is certainly worthwhile to have a context for classification, but this should be applied to Western classification with the same purpose. Operating with the understanding of different worldviews means that there can be no "mistakes" in classification. As Gurung cautions: "Taking the Linnaean system as a point of reference might do injustice to the qualities of folk-classifications, as it unconsciously includes the assumption that modern systems are more advanced and meaningful because they allow scientists from different parts of the globe to exchange information based on exactly the same principles and rules. It remains to be questioned, however, whether such a system is natural and the most appropriate for all" (2003, 365).

Examining the classification of social insects through historical changes, gendered perspectives, and alternative cross-cultural terms and analogies illustrates the ways in which this classification is socially constructed. Classification is a social endeavor and reflects the social and political context of those who do the classifying. The historical and cross-cultural examples provided show the specific Western classification of social insects to be a social construct. This construct does not simply serve as a convenience of language nor does it represent universal interpretation. Breaking down the hegemonic assumption that the classification of social insects is best reflected by the Western model serves as a first step in developing alternative models.

4

Entomologists and Sociologists
COMMON GROUND

When it comes to general interactions between biology and the social sciences, the received view tends to be that the social sciences initially borrowed concepts and methods directly from biology in order to gain the validation and legitimacy attached to this discipline. Sociohistorical studies of the use of theories, analogies, and concepts therefore tend to focus on the transfer from biology to the social sciences. However, by closely examining the interactions of specific disciplines, a more complex view emerges that may reveal a mutual exchange of ideas. I contend that this is true of sociology and entomology during the early formative history of their respective disciplines in the nineteenth and early twentieth centuries. Their mutual exchange of theories and concepts enhanced the emerging fields' credibility and were tied into larger evolutionary and ecological paradigms. Although examples of social insects were certainly used to "naturalize" ideas of human social organization, the concepts and terminology of human sociality were and still are used to explain social insect behavior and organizational structure. In examining the interactions between sociology and entomology, we can form a better idea as to the overall impact their mutual exchange of ideas has had on the study of both human and social insect behavior and organizational structure.

SIGNIFICANCE OF INTERACTIONS

Much has been written about the effect of biology on the social sciences. However, shared theories and concepts in the nineteenth and early twentieth centuries did not just leave a mark on the social sciences in general; biological analogies also involved the interpretation of the social writ onto the natural world. The term "survival of the fittest" is an apt example. Herbert Spencer coined this term, and Darwin used it in *The Origin of Species* (Greene 1981; Paul 1988; Sapp 1994). Darwin's idea of natural selection is argued to have come directly from Thomas Malthus's *Essays on Population* (Becker and Barnes 1961; Mayr 1977; Thomson 1998; Vorzimmer 1969; Young 1969).

In fact, evolutionary theories and the attending concepts became a common thread between the natural and social sciences. Before the boundaries between disciplines became firm, there was a cross-fertilization of ideas concerning social organization and the development of societies. As Jones observes: "This conviction of the possibility of cross-disciplinary general laws and of the close connection between biology and sociology supported the transfer of concepts and ideas between them" (1980, 3). Ecological principles became another theoretical connection between the social and the natural sciences (Gross 2002; Wali 1999; Worster 1985). Gross argues that ecological theories were formed out of the mutual interactions between biologists and sociologists. In particular, he examines the interrelationship of the sociologist E. A. Ross, the plant ecologist Frederic E. Clements, and the Ecology Group at the University of Chicago to support his contention. Gross notes that sociologists of the late nineteenth and early twentieth centuries frequently cited the work of animal and plant ecologists. But he also points out that the prominent plant ecologist Clements was heavily influenced by sociologists and in particular Ross (2002, 2–3). It would seem that both evolutionary theories and ecological theories provoked interactions between various disciplines within the natural and social sciences.

Limoges investigates the complex interactions between the social sciences and the natural sciences by way of the division of labor concept. One of his premises is that not only are there mutual exchanges but there may be times when the natural sciences benefit more from these exchanges than the social sciences (1994, 317). And Ogburn and Goldenweiser determined that, "biology has benefited the social sciences by introducing the concept of natural growth and by defining the scope and limits of man's organic traits; but . . . it has also hurt them by flooding the field of social theory with the dogmatic notion of a rigidly ordered development (1927, 7).

Limoges, as well as Ogburn and Goldenweiser, highlight the mutual nature of exchanges. However, they contradict the received wisdom by also indicating that natural science may have gained more from these exchanges than have the social sciences. This may not always be true for specific disciplines within the natural or social sciences; more detailed analyses of various interactions may illuminate how each fared. I. B. Cohen (1994) insists that not only do general interactions between the natural sciences and the social sciences need to be discussed, but that the interaction between specific disciplines must begin to be examined. This has proved fruitful for economics, as Phillip Mirowski's exemplary edited volume *Natural Images in Economic Thought: "Markets Read in Tooth and Claw"* attests to specific exchanges that reinforced both economic theory and ideas about the natural world. Examining the interactions between entomology and sociology reveals a mutuality of theories, analogies, and concepts as well as how this mutuality shaped each discipline initially and how it continues to have current implications for the respective fields.

THE NINETEENTH-CENTURY
CONTEXT—EVOLUTIONARY THEORIES

One of the difficulties in discussing early interactions between entomologists and sociologists comes from the fact that previous to the establishment of fixed disciplinary boundaries, the training of scientists was broad and they wrote on both biological and sociological topics. Charles Darwin, Alfred Russel Wallace, Herbert Spencer, and Peter Kropotkin, to name a few, wrote on both human and animal societies, fueled by their backgrounds, interests, and belief in general laws that would apply to both realms. I impose rough divisions between the "social" and the "natural" in these earlier cases in order to highlight the exchange of ideas emerging from broadly defined sociological and entomological contributions. In the emerging fields of sociology and entomology, these contributions occurred within a larger context of evolutionary theories. Sorenson maintains that evolutionary theory had special import for entomologists because they identified with Darwin as a fellow entomologist. Darwin was, after all, one of the founders and a lifetime member of the Entomological Society of London. Sorenson lists Darwin's other related credentials and states that a "considerable portion of his published work dealt with insects" (1995, 197). He especially points to *The Descent of*

Man (1871) as a work that extensively relies on entomological data, as did Darwin's ideas for evolution in general (1995, 197–98). The entomologist Herbert H. Ross concurs with this and claims that both Darwin and Wallace "were influenced by evidence from insects" for their formulation of natural selection (Ross 1973, 172).

Clearly, evolution was not of interest only to entomologists and other natural scientists but was important for the social sciences and sociology as well. Darwinian theory proposed continuity between humans and other animals and put forth the idea of natural selection as key to evolutionary progress. However, ideas of biological evolution and social evolution had been developed previous to Darwin and Wallace, and therefore Darwinian theory was a part of a preexisting exchange of ideas that went beyond the natural sciences. As Jones points out: "Darwinism shared a vocabulary with social thought. It was, therefore, 'recognisable' to social theorists in certain specific ways and had links with a much broader culture than other nineteenth-century scientific developments" (1980, 5). Considering this, the exchange between sociology and entomology had occurred in the development of the idea of evolution. Darwin and Wallace utilized observations of the natural world as evidence to generalize social ideas that they knew of from the political economy of Malthus, as well as the sociological work of Herbert Spencer and others, including the term "evolution." Robert Park remarked on the influence of sociology on Darwin specifically: "It is interesting to note that it was the application to organic life of a sociological principle—the principle, namely, of 'competitive co-operation'—that gave Darwin the first clue to the formulation of his theory of evolution" (1936, 2–3).

The influence of Herbert Spencer and his more Lamarckian evolutionary ideas was significant not only for Darwin's development of evolutionary theory, but for all of natural and social science. Because Spencer is most known in connection with social Darwinism, there has been an overemphasis placed on his transference of biological precepts onto the social world. In fact, he was highly influential in both the natural and the social sciences at the time. As Franklin Giddings claims in discussing the origin of the words "sociology" and "biology": "'[B]iology,' like 'sociology,' had no vogue until Mr. Spencer took it up" (1896, 32). Spencer's ideas concerning the natural and social sciences also fostered exchanges between specific disciplines. Spencer's influence on entomology can be seen in the president's address given by Alfred Russel Wallace to the Entomological Society of London in 1872. Chastising his fellow entomologists for overlooking Spencer's contribution during their ongoing discussion on insect evolution, Wallace admonished the group: "[I]t is to me surprising that one of the most ingenious and remarkable theories ever put forth on a question of natural history has not been so much as once alluded to. More than six years ago, Mr. Herbert Spencer published . . . a view of the nature and origin of the annulose type of animals, which goes to the very root of the whole question; and . . . it must so materially affect the interpretation of all embryological and anatomical facts bearing on this great subject, that those who work in ignorance of it can hardly hope to arrive at true results" (Wallace 1872, 350).

Wallace then continued this speech, placing Spencer's theory concerning insect evolution within a larger context of ideas and importance that would thereby create general laws. He asserted that Spencer's theory "may throw light on many an obscure problem, and which will perhaps materially influence our ideas as to the nature of life itself (1872, 351). Wallace viewed Spencer's theorizing about the origin of insects as connected to natural laws that could pertain to human society as well. At the point that Wallace gave this address, he was much impressed with Spencer, even naming one of his sons Herbert Spencer Wallace (Jones 1980). If he was convinced of Spencer's positive contribution to entomology and larger general laws, he later fell away from Spencer over more definite sociological issues, as did other followers of Spencer. One of these other followers was Grant Allen, who claims that disagreements over socialism explain the disenchantment with Spencer that developed among the group that was once considered his "disciples": "The rock on which he split with his younger disciples was Socialism. Very early, most of those whom he had profoundly influenced had been led by the perusal of 'Social Statics' into the acceptance of his original idea of Land Nationalization. Alfred Russel Wallace, the chief English exponent of the doctrine, founded his argument entirely on Spencer. Later on Wallace became a convinced Socialist, as did most of the other thinkers whose opinions Spencer had most deeply leavened" (1904, 626). Spencer himself repudiated any earlier ideas that may have appeared sympathetic to socialism.

Allen's account makes it clear how difficult it is to separate out the exchanges between those in the sociological and biological fields. Wallace, president of the London Entomological Society and "co-discoverer" of Darwinian evolution theory, was also engaged in sociological issues and involved in a group that followed Spencer for his earlier sociological writings. As the quotation above suggests, combining these fields was not unique to Wallace, who felt that he had made contributions to the "science of sociology" as well as the obvious biological contributions (Fichman 1997). Likewise, Spencer's influence on theories of sociology and entomology was widespread during the nineteenth century. The fact that Lamarckian theories of inheritance were embraced by many social theorists as well as biologists at the time contributed to the popularity of Spencer (Jones 1980; Kingsland 2005).

Lamarckianism even crept back into Darwin's later work, yet it was only because he could not fully establish the cause of heredity at that time. Whereas this caused problems for Darwin's theory, the explanation of heredity through the use-inheritance of Lamarckianism "seemed to solve so many problems of social theory. It united biology with sociology. It explained how evolutionary change and social behaviour were linked" (Jones 1980, 79). Spencer particularly found the connection, "the best guarantee that behaviour and society could be rapidly brought into correspondence with one another" (Jones 1980, 80). Clearly, Spencer's ideas of evolution provoked an interchange between the fields of sociology and biology. More examples of Spencer's influence on the specific disciplines of sociology and entomology will be provided in the discussion of shared concepts. Before moving on, however, another approach to nineteenth-century evolutionary theory must be presented.

Kropotkin and "Darwin without Malthus"

Evolutionary theory of the nineteenth and early twentieth centuries was not monolithic, which presents an even more complex picture of the interactions between the natural sciences and social sciences. That Darwin utilized Malthus's ideas on human population and that this may have biased his theory toward competition in nature is not a recent realization discovered with the luxury of historical hindsight; critics of the day also mentioned this potential problem. In fact, Cronin explains that the extent of the critique could be considered an "alternative tradition" that stressed cooperation over competition (1991, 269).

The Russian natural scientist and sociologist Kropotkin was one of the earliest to reject the Malthusian overtones of Darwin's work. Cronin remarks that Kropotkin and others with similar views were "explicitly dissenting" from what they knew to be the dominant discourse (1991, 269). Kropotkin was writing not just as a rebuttal to Darwin himself, but more to Darwin's followers, whom he accused of oversimplifying the findings of Darwin and focusing on only one kind of "survival of the fittest." He felt that Darwin was aware of a larger sense of survival that included symbiotic relationships but that his followers ignored this larger sense of interrelationships. Kropotkin understood this to be not only a limiting and distorted development for natural sciences but also for social sciences, specifically because these were intertwined and came to share the same underlying assumptions about evolution.

> The conception of struggle for existence as a factor of evolution, introduced into science by Darwin and Wallace, has permitted us to embrace an immensely wide range of phenomena in one single generalization, which soon became the very basis of our philosophical, biological, and sociological speculations. An immense variety of facts:— adaptations of function and structure of organic beings to their surroundings; physiological and anatomical evolution; intellectual progress, and moral development itself, which we formerly used to explain by so many different causes, were embodied by Darwin in one general conception. (1902, 1)

Sociologists and entomologists who followed Kropotkin placed emphasis on cooperation in their views of human and insect society and those who followed Darwin tended to use examples that supported survival of the fittest or competition on individual terms. Although recognizing the concept of "survival of the fittest," Kropotkin nevertheless maintained that this was accomplished on the group, not the individual, level. The fact that the presence of altruism was a difficult problem to be solved for Darwin, whereas for Kropotkin it was a given factor of survival, points to their fundamental differences in approaching the social and natural world. Kropotkin used social insect evidence in a different way than did Darwin or his followers in order to exhibit a general law of cooperation for human and insect societies: "If we take an ant's nest, we not only see that every description of work—rearing of progeny, foraging, building, rearing of aphids, and so on—is performed according to the principles of voluntary mutual aid" (1902, 12).

Kropotkin specifically aligned himself within a larger group of scientists who did not accept competition as the main factor in natural selection. He discussed the Russian zoologist Kessler's refutation of competition as an overriding principle and maintained that most other Russian zoologists rejected the importance of competition as well, based on their fieldwork observations (1902, 7–10). Kropotkin's rejection of the social interpretation of competition in the natural world is summed up in his statement, "The ants and termites have renounced the 'Hobbesian war,' and they are the better for it" (1902, 14).

Kropotkin had an influence in the wider literature outside of Russian science as well. In veiled terms, Kropotkin indicated that the published ideas leading up to his book *Mutual Aid* appear in later sociological works.

> Of works dealing with nearly the same subject, which have been published since the publication of my articles on Mutual Aid among Animals, I must mention The Lowell Lectures on the Ascent of Man, by Henry Drummond (London, 1894), and The Origin and Growth of the Moral Instinct, by A. Sutherland (London, 1898). . . . A third work dealing with man and written on similar lines is The Principles of Sociology, by Prof. F. A. Giddings, the first edition of which was published in 1896 at New York and London, and the leading ideas of which were sketched by the author in a pamphlet in 1894. I must leave, however, to literary critics the task of discussing the points of contact, resemblance, or divergence between these works and mine. (1902, xviii)

In fact, Franklin Giddings cites Kropotkin and what he calls his "remarkable papers," in his *Principles of Sociology* and uses his entomological examples to make a case for cooperation as integral to the social cohesion of human society. Giddings seems convinced by Kropotkin's extensive analogies illustrating cooperation. Summing up Kropotkin's main emphasis on mutual aid, Giddings sides with him that cooperation is the more important principle for life in societies, be they human or animal:

That life in societies is the most powerful weapon in the struggle for life, taken in its widest sense, has been illustrated by several examples on the foregoing pages, and could be illustrated by any amount of evidence, if further evidence were required. Life in societies enables the feeblest insects, the feeblest birds, and the feeblest mammals to resist, or to protect themselves from the most terrible birds, and beasts of prey; it permits longevity; it enables the species to rear its progeny with the least waste of energy and to maintain its numbers albeit a very slow birth rate; it enables the gregarious animals to migrate in search of new abodes. Therefore, while fully admitting that force, swiftness, protective colours, cunningness, and endurance to hunger and cold, which are mentioned by Darwin and Wallace, are so many qualities making the individual, or the species, the fittest under certain circumstances, we maintain that under any circumstances sociability is the greatest advantage in the struggle for life. (1896, 205–6)

The interactions between prominent sociologists and entomologists provided thus far are representative of the impact that evolutionary theories had on these fields in general, and how entomologists and sociologists utilized each other's ideas concerning evolution, thus shaping the debate on the origins and development of human and insect societies. Specific shared concepts that emerged from this ongoing interchange of ideas will be discussed later. For now, I turn to another broad theoretical approach toward analyzing societies that was taken up by both sociologists and entomologists, fostering yet another set of interactions that had implications for both fields.

THE EARLY TWENTIETH-CENTURY CONTEXT—ECOLOGICAL THEORIES

The interactions between the social sciences and biological sciences over evolutionary theories help to explain much of the foundation for shared biological analogies and conceptions in the nineteenth and early twentieth centuries. However, so does the later interaction between natural scientists and social scientists in the field of ecology. Sorenson (1995) notes of U.S. entomologists in particular that their early acceptance of evolutionary principles led them to an interest in ecological principles. He contends that, "by the 1880s, the Americans comprised the largest and most productive group of practicing ecological investigators in the world" (1995, 256). These ecological principles would be another intersection with sociology and entomology that developed and would hold sway until the 1940s.

Many social and natural scientists who embraced ecological thinking were involved in the connections of the social and the natural because of their belief that ecological principles were general laws that would apply to both fields. Therefore, ecological approaches were taken up by sociologists and biologists for very similar reasons. Worster names two significant reasons for the ecological turn. Ecological theories were a reaction against both a mechanistic and individualistic orientation that had begun to develop in the sciences. Lewis Mumford, Robert Park, and the Chicago school wanted to apply ecology to understand communities. Ecologists such as Patrick Geddes, C. C. Adams, and Walter Taylor felt that ecological insights disproved the existing emphasis on individual competitiveness in nature (Worster 1985, 319–20). Ecology offered a united front for those from the natural sciences and social sciences who would challenge the focus on individualism that had been pervading both fields. Crossing the emerging disciplinary boundaries reinforced and gave credibility to each through the recognition of familiar terrains.

Mohan Wali (1999, 41) credits Patrick Geddes with being the first one who "used the term ecology in an interdisciplinary context in 1880 in his illustration of the proper placement of disciplines." Geddes influenced sociologists such as Lewis Mumford and Radhakamal Mukerjee in their use of social ecology. The interpretation and application of general ecological laws pertaining to humans and animals remained a connection for entomologists and sociologists specifically. Worster (1985) notes that the prominent entomologist William Morton Wheeler was a vocal proponent of organistic ecology during the first half of the twentieth century, attributing Wheeler's adoption of this philosophy to his study of ants and the model they provided (1985, 320). Wheeler was extremely well read in many fields outside of his area, and he utilized this wide range of knowledge in his entomological studies (Parker 1938). He felt that the lessons provided by social insects crossed over disciplinary boundaries: "I believe that the study of the social insects has, at the present time, a peculiar interest to the serious student of philosophy, sociology, and animal behaviour. Since we ourselves are social animals—I had almost said social insects—the philosophically inclined cannot fail to find food for thought in the strange analogies to human society, which continually reveal themselves among the wasps, bees, ants and termites, and the behaviourist will note that they suggest a bewildering array of fascinating facts and problems" (Wheeler 1928a, 1–2).

Wheeler was particularly interested in the discussion of instinct and social insects as it appeared in sociological literature as this was a connection between the natural and social sciences. He felt that the "very elaborate social behaviour of the insects" became theoretically important in that it offered an example of instinctual behavior for human social theory. "We are beginning to see that our social as well as our individual behavior is determined by a great background of irrational, subconscious, physiological processes. Any doubts in regard to the existence of this substratum will be dispelled by a perusal of Pareto's 'Treatise of General Sociology' (1917), the first volume of which is devoted to these 'residues' which condition our social activities" (Wheeler 1928a, 2).

Charlotte Sleigh has examined the significance of Pareto for Wheeler's work, especially for his idea on trophallaxis as a means of social cohesion (2002). She describes Wheeler's reaction to Pareto's work as marked by "proselytic fervor" (2002, 150) and believes that both Aldous Huxley and Wheeler shared "a deep debt to Pareto" for their views on behavior and human nature (2002, 152). Beyond this important connection to Pareto's work, Wheeler also credited other sociologists and their views on social cohesion. What he seemed to admire about all of these sociologists' views is that they embraced a nonrationalistic approach to social cohesion. "The very significant rôle of the primitively psychological and the relative insignificance, even in our present civilization, of the specifically intellectual processes have been most impressively set forth by Pareto in his 'Traité de Sociologie' and by Sumner and Keller in their 'Science of Society.'. . . Pareto designates the irrational foundations of social behavior as the 'residues.'. . . Sumner and Keller's remarkable picture of mores . . . forms an admirable background for Pareto's contentions, which he illustrates mainly with materials drawn from the ancient and contemporary history of European peoples" (Wheeler 1930, 154).

Wheeler refers to a trend of examining nonrational behaviors; this was part of a larger debate on instinct that intersected with evolutionary and ecological theories and involved both entomologists and sociologists. I discuss this debate in depth in chapter 5; for the present, it is enough to realize that analogies between human and insect societies allowed for exchange of ecologically based evidence, theories, and concepts between entomologists and sociologists. Wheeler's collaborative efforts, such as creating a course on comparative sociology with Pitrim Sorokin at Harvard or being influenced by Pareto and others, are a testament to this connection (Sleigh 2002). Wheeler frequently cited other sociologists in his work, such as Comte, DeGrange, Spencer, and Durkheim, on ideas of organicism, accumulation, social cohesion, division of labor, and social parasitism, and in some cases their insights helped him to form certain concepts that were then transferred back to sociology with the "natural" evidence from entomology to further support their credibility. And Wheeler was not alone in this, as Sleigh points out: "Ant–human metaphors . . . [were] freely exchanged between sociologists and entomologists in the early twentieth century" (2002, 155).

The University of Chicago Ecology Group

Another set of interactions between the natural and social sciences occurred at the University of Chicago and centered on the development of ecological ideas to describe nonhuman and human societies. The interdisciplinary Ecology Group was prominent in the 1930s and 1940s, and their work had wide influence. "During the 1930s and -40s this cluster of scientists gathered informally every other Monday evening in the parlor of Professor Warder Allee to share their findings and insights" (Worster 1985, 326). W. C. Allee published his classic *Animal Aggregations: A Study in General Sociology* in 1931 and viewed his work as having implications for sociology and biology. Allee, who published in the *American Journal of Sociology* and was influenced by the sociological work of Alfred Espinas and Peter Kropotkin, credits his conception of "proto-cooperation" as emerging from an original idea of Alfred Espinas (Collias 1991, 618). In his quest to present a cooperative view of sociality, Allee placed humans and insects on a parallel evolutionary path that traced their common origins of sociality back to the protozoa (Mitman 1992) (see fig. 2).

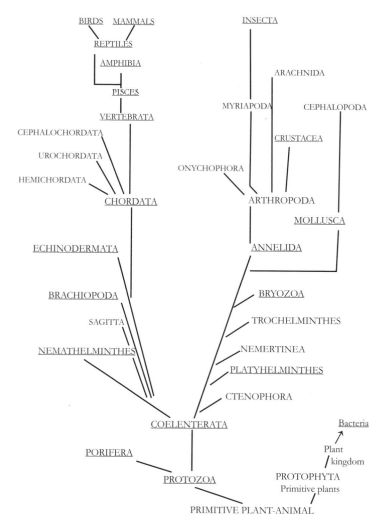

FIGURE 2. W. C. Allee's Evolutionary Comparison. Phyla are shown in larger font, classes in smaller. Redrawn from W. C. Alee, 1931, "Co-Operation Among Animals," *The American Journal of Sociology* 37: 386–98.

Although most well-known for his work on dominance hierarchies in certain vertebrates, Allee's focus was on the more frequent cooperation found in the rest of the animal kingdom. He saw cooperation as the goal to strive for in society and the key to social cohesion. His ally in this endeavor was Alfred E. Emerson, entomologist and Allee's coauthor of the classic *Principles of Animal Ecology* along with Thomas Park, Karl Schmidt, and Orlando Park. Emerson shared Allee's concern with cooperation and social cohesion and sought to utilize scientific evidence to advance cooperative organization in human society. Despite any differences between the two fields, Emerson felt that sociology and entomology should share comparisons to produce important underlying principles of social organization. His 1942 article "Basic Comparisons of Human and Insect Societies" provided his reasoning for the benefits of a working relationship between entomologists and sociologists:

> The development of human social heredity through learned symbols is of such importance that this human attribute would seem to indicate the valid division line between the social and the biological sciences. It is what the sociologist means when he says that man is unique in the possession of culture. However, even though techniques may differ and phenomena are diverse in numerous instances, I personally believe that scientific method is fundamentally the same whether applied to human social mechanisms or insect social mechanism or whether applied to the social supraorganism or to the individual organism. In spite of the real differences between the societal types, careful comparisons and correlation leads us to the formulation of important principles. (Emerson 1942, 169)

Emerson argued that conscious interaction across disciplinary lines would be helpful in many ways. Not only would it help to uncover universal laws, but it would also aid in the prediction and application of these laws. Principles of sociality would then become a cumulative body of knowledge. Addressing these benefits, Emerson posed the question concerning a correlation between insect and human societies and explained the advantages: "Significant relationships between diverse phenomena indicate more universal and thus more fundamental principles thereby enabling us to arrange social attributes in a more logical and scientific order. Chronological order may help us to detect causative factors more easily and to predict events with greater accuracy. Social perspective and understanding will grow. Social control based on the practical application of our knowledge of social mechanisms will be nearer. For these reasons, I recommend a closer reciprocal understanding between the social and biological sciences, especially when correlated phenomena are under investigation" (1942, 175).

Emerson's and Allee's interest in cooperative social organization in termites and other animals was informed by their concern with social issues. As their careers spanned both world wars, they coped with the realities of war and the rise of fascism (Mitman 1988). Allee and Emerson became engaged in discussions, as did the larger scientific community, on the role science would play in war or, alternately, in efforts to bring about peace. They were more inclined toward peace efforts, and Allee in particular was a part of the early peace movement; "both Allee and Emerson saw a need for a world order based on a greater degree of integration and cooperation" (Mitman 1988, 190).

The sociologist Robert E. Park was also involved in the Ecology Group and contributed to the interdisciplinary volume of *Biological Symposia VIII*, edited by the anthropologist Robert Redfield, who credited Park with introducing fellow social scientists to contemporary sociological ideas (Martindale 1960, 93). Park's article "Human Ecology" was published in the *American Journal of Sociology* in 1936. In it, he described human ecology as a new area of research that was "an attempt to apply to the interrelations of human beings a type of analysis previously applied to the interrelations of plants and animals" (1). Park identified human interactions as a part of the larger "web of life," although noting the differences between human groups and plant or animal groups. In citing the work of natural sciences, he elaborated on what made human ecology unique. For instance, although he found *The Science of Life* written by H. G. Wells, Julian Huxley, and G. P. Wells to be interesting, he still felt that the pressures and circumstances of human society were different from those facing animal society. Park found that the example of social insects in the work of the organic ecologist William Morton Wheeler helped to explain the necessary distinction between these two groups: "In plant and animal communities structure is biologically determined, and so far as any division of labor exists at all it has a physiological and instinctive basis. The social insects afford a conspicuous example of this fact, and one interest in studying their habits, as Wheeler points out, is that they show the extent to which social organization can be developed on a purely physiological and instinctive basis, as is the case among human beings in the natural as distinguished from the institutional family" (Park 1936, 13). Granting a purely instinctive foundation for human individuals similar to that of social insects, Park elaborated on the layer that differentiated humans as a group from social insect societies. Human societies underwent institutionalization, and further, "[i]n human as contrasted with animal societies, competition and the freedom of the individual is limited on every level above the biotic by custom and consensus (13). Park's position on this was consistent with the idea of integrative levels promoted by the Ecology Group. Each level was subject to ecological laws that would become increasingly complex, and these levels could be compared in some ways but not others.

The interdisciplinary scientists of the Ecology Group at the University of Chicago were not the only ones who utilized ecological principles. Many other researchers such as C. C. Adams and Charles Elton were advancing the ecological framework. The sociologist Edward A. Ross at Wisconsin also adopted the new ecological perspective. Gross (2002) maintains that while Ross differed from Robert Park as to the amount of social reform that should be sought in society, the basic ecological principles of social organization were important to both sociologists. Another connection between them was the Indian sociologist Radhakamal Mukerjee, whom both visited in India and maintained contact with over the years. Robert Park provided space for Mukerjee to study in Chicago, and Ross published and wrote the preface to Mukerjee's *Regional Sociology*, a book that presented the ecology of regionalism and used a liberal dose of comparisons to animal societies, including social insects (Gross 2002; Mukerjee 1926). The work of William Morton Wheeler was cited frequently in Mukerjee's writing, and while embracing ecological organicism Mukerjee also revived the emphasis on mutual aid in evolution as proposed by Kropotkin. Mukerjee went so far as to proclaim: "It is well known that the emphasis of struggle in Darwinism is responsible for a one-sided interpretation of man's behaviour in the past century which has now been considered misleading" (1940, 8). Rejecting the struggle for existence at the individual level corresponded to Mukerjee's idea of symbiotic relationships within regions. Similar to Kropotkin and Russian zoologists, Mukerjee used his knowledge of the natural and social systems in India to advance his ideas of regionalism and symbiosis. Sociologists and entomologists exchanged ideas and shared in overarching theories that encouraged mutual interaction throughout the nineteenth and early twentieth centuries. Within the context of evolution and ecology, analogies comparing human and social insect societies included a shared lexicon. Many of the concepts that were developed remained within the vocabulary of both fields.

SHARED CONCEPTS

Coming from the larger contexts of evolutionary and ecological theories as they applied to their respective fields, sociology and entomology easily exchanged and shared concepts used to describe social organization and behavior. This exchange of concepts and analogies mutually reinforced the credibility of both emerging disciplines. Entomology provided the legitimacy of naturalized examples for sociology, whereas sociology provided preexisting concepts from human societal analogies that entomologists then used to describe insect behavior and structure. Together they created an exchange of concepts that operated under the premise of general laws that were assumed to apply for both human and insect societies, whether those were evolutionary laws or ecological ones. However, this sharing of concepts may have had dissimilar effects for entomology and sociology as each discipline developed.

Limoges (1994) calls for a more critical analysis of concept transfer than has previously been given in the history of interrelationships between the natural and the social sciences. He insists that concept transfer cannot be considered "simple borrowings or isomorphic processes" (1994, 318). By closely analyzing a transfer between Milne-Edwards, Darwin, and Durkheim, Limoges concludes that in this case, as in others, the transfer of concepts cannot be considered a simple "univocal process" (1994, 335). Instead, "[c]onceptual transfers may have very different functions, not necessarily mutually exclusive. They may serve as heuristic and theoretical construction tools, as pedagogical and persuasion devices, as polemical weapons, as legitimizing labels, or as evidential support" (1994, 336). The conceptual transfers that occurred between entomology and sociology undoubtedly fall under all of these categories. Shared concepts describing human and social insect societies came from shared organizational assumptions. I present interactions concerning three very important concepts for both fields: division of labor, caste, and superorganism.

Division of Labor

Darwin's ideas of natural selection and division of labor were initially influenced by the social sciences, so it is no surprise that he found evidence of the same in the natural world, specifically in his study of social insects:

> I have now explained how, as I believe, the wonderful fact of two distinctly defined castes of sterile workers existing in the same nest, both widely different from each other and from their parents, has originated. We can see how useful their production may have been to a social community of ants, on the same principle that the division of labour is useful to civilised man. Ants, however, work by inherited instincts and by inherited organs or tools, whilst man works by acquired knowledge and manufactured instruments. But I must confess, that, with all my faith in natural selection, I should never have anticipated that this principle could have been efficient in so high a degree, had not the case of these neuter insects led me to this conclusion.
> (Darwin [1859] 1998, 358)

Having previous knowledge of the division of labor in human society and an idea of natural selection derived from the social sciences, Darwin then "discovered" this same general law in the natural world of the social insects. The convergence between entomological and sociological evidence produced a simultaneous legitimacy, reinforcing certain ideas about the natural and the social at the same time and trading off of the familiarity of both realms. This process of legitimation did not go unnoticed at the time either. In a letter to Engels in 1862, Marx wrote: "It is remarkable how Darwin rediscovers, among the beasts and plants, the society of England with its division of labour, competition, opening up of new markets, 'inventions' and Malthusian 'struggle for existence.'. . . [I]n Darwin, the animal kingdom figures as civil society" (Marx and Engels [1862] 1986, 381). Whereas Darwin saw the division of labor as a product of natural selection, useful and efficient for both ants and men, Marx and Engels recognized the naturalizing of nineteenth-century political economy in this perception.

Kropotkin also criticized the idea of social Darwinism as a distortion of evolution and politically motivated bias. He held a less rigid view of the division of labor and used alternative instances of entomological evidence to illustrate this: "Former observers often spoke of kings, queens, managers, and so on; but since Huber and Forel have published their minute observations, no doubt is possible as to the free scope left for every individual's initiative in whatever the ants do" (Kropotkin 1902, 32). Mutual aid for Kropotkin involved working together, but not in any permanent specialization of roles or tasks. In support of the anarchist-communism that he espoused, Kropotkin believed that the rigid division of labor found in capitalism was not found in nature. He found a particular example of this with bees: "By working in common they multiply their individual forces; by resorting to a temporary division of labour combined with the capacity of each bee to perform every kind of work when required, they attain such a degree of well-being and safety as no isolated animal can ever expect to achieve however strong or well armed it may be. In their combinations they are often more successful than man, when he neglects to take advantage of a well-planned mutual assistance" (1902, 16).

The division of labor is a theoretical concept that is extremely important to both sociology and entomology. Division of labor in entomology is the main characteristic that determines the highest level of sociality in insect organization as exhibited in those insects termed "eusocial." Eusocial insects are found in certain species of ants, bees, termites, and wasps exclusively. For sociology, division of labor has also been associated with the degree of complexity in social organization. The sociologist Emile Durkheim, known for his classic work *The Division of Labor* (1893), credited the zoologist Milne-Edwards for the concept of physiological division of labor and also Adam Smith for the division of labor concept as applied to human society (26). Durkheim posited that mechanical solidarity came about through a simple division of labor and that the organic solidarity found in modern society emerged through a highly specialized division of labor. Division of labor was the key to social cohesion for Durkheim, and this idea held fast in sociology, influencing the direction of inquiry especially in regard to organizational structure.

In his work on social insects, the entomologist William Morton Wheeler included Emile Durkheim as one of the sociologists he admired in regard to theories of social cohesion. While the theories of sociologists such as Comte and Spencer seemed limited and narrow to Wheeler, Durkheim's theory escaped this critique, as his view appeared to be the most universal, allowing Wheeler to apply it to animal societies. "Durkheim's view has the advantage of referring the integration, or solidarity of society, to a principle which is universal, not only in animal societies, but also in all multicellular organisms. This principle, the division of labor, was first recognized and named by economist Adam Smith and only later introduced into biology by Milne-Edwards" (Wheeler 1930, 153–54). Wheeler also very clearly established the source of the term in the social sciences rather than biology after claiming its universality. He may have understood this connection through Durkheim's own acknowledgment of its sources. Division of labor as the source of social cohesion seems to be adopted by Wheeler through his agreement with Durkheim. Taking into consideration that Wheeler was a prominent myrmecologist who was building definitions for insect sociality, this concept transfer is extremely significant. Wheeler shared with Durkheim a sense that the highest level of social development occurred with a highly specialized division of labor. Believing the division of labor to be a concept shared by humans and insects, Wheeler helped to establish a ranking of sociality where the highest ranking for insects would be marked by a specialized division of labor. Wheeler emphasized extended parental care and the reproductive division of labor for eusociality while the revised definitions of the later authors Michener and Michener (1951) and especially E. O. Wilson would expand on this and claim division of labor more generally (Hölldobler and Wilson 1990). As the current definition stands, the role of division of labor in determining eusociality is far greater than Wheeler's initial use of a reproductive division of labor.

Although a bit out of the boundaries of the time period discussed, I include the following statements from Michener and Michener's *American Social Insects* published in 1951 to illustrate the assumption that justified the sharing of the division of labor concept between entomologists and sociologists. "Only man and certain insects have achieved true societies in the sense that only they have gone on to employ division of labor" (240). In comparisons then and now, this concept was understood to vary in its expression within human or insect societies, despite its marker of highly complex sociality for both. "Thus division of labor among insects is based on physical differences between individuals. In man it is based largely on mental differences and differences in training" (242).

In the work of the sociologist Radhakamal Mukerjee, there appears what would become a familiar theme of these sorts of comparisons: that insects and humans achieve a specialized division of labor by different means. Mukerjee also seemed to attribute to social insects a "perfected organization" due to the extreme level of specialized division of labor that resulted in polymorphism. "Sociable like man, social insects, by this means, have achieved the harmony and effectiveness of a perfected organization which seems to realize the dreams of Utopians. In the process of survival, they have undergone an evolution parallel to man's; but, while polymorphism is the basis of mutual service and division of labor among insects, man has achieved the same result by his intelligence and social consciousness" (1926, 227).

Division of labor as a shared concept privileges a particular type of organization for both social insects and humans. Both disciplines have stressed that organizational forms with the most specialized division of labor are the most complex. These organizational forms with a highly specialized division of labor are also linked with concepts based on hierarchy and caste, to the latter of which I now turn.

Caste

The idea of caste fostered a connection between entomological and sociological literature because of its role in relation to a highly specialized and rigid division of labor. This view of social organization, whether human or insect, became legitimated as a reflection of structural reality through an exchange of the concepts between the disciplines. To reinforce the idea of caste, entomologists and sociologists had only to refer to the natural or the social analogy, which then in turn generalized the concept and made it appear familiar.

The caste system in India and the interpretation of caste systems in the organization of social insects became analogies that in the late nineteenth and early twentieth centuries began to be used to explain racial stratification, especially in the United States. Sociologists were influenced during this period by ideas of heredity and natural selection and felt that some groups would not progress but rather were destined to occupy a lower station in life (Williams 1989). A "school" of caste developed at the University of Chicago in the 1930s and 1940s, primarily through the work of Lloyd Warner and the social scientists who followed him. The sociologist Oliver Cox criticized this school for its inaccurate conflation of race with caste. Although he credited some other sociologists for beginning to recognize the problems associated with conflating race with caste, Cox pointed out that Robert Park and others included the notion of caste in their textbooks and employed the concept to describe race relations in the South (1942).

Robert E. Park provides an interesting case for the eventual shift within sociology of the use of this term. Park's ecological perspective caused him to attribute more weight to the environmental explanations of caste than to those that emphasized biological heredity. However, this was still a very deterministic position on race. He developed a theory of race relations in 1928 that described a social system of "biracial organization" that included notions of caste. Although not advocating a caste system as an ideal end result, Park seemed to believe that there were obstacles that prevented the assimilation of blacks in the United States. The "accommodation" stage that seemed more realistic to Park at the time involved a "stable but unequal social order, such as a caste system" (Williams 1989, 143). Over time, Park moved from this position to acknowledge in 1937 that blacks were not a caste but rather a minority moving toward full assimilation (Williams 1989). Despite Park's move beyond caste and critics such as Oliver Cox, theories of caste were still prevalent in the general sociological literature until the late 1940s.

On this issue, Mukerjee may have been influenced by the University of Chicago Ecology Group, and Robert E. Park specifically, which would explain his adoption of the caste notion to describe race relations in America. Although he critiqued Western descriptions of the caste system in India, something that Cox had done as well, he then proceeded to suggest that a caste system should be *correctly* instituted in the South to ease race relations. His solution for race relations in the South was reflective of Park's earlier concept of biracial organization. Mukerjee's work is also a good example of the frequent and coexistent use of entomological and sociological evidence for caste systems that existed in both the entomological and sociological literature of the time. Mukerjee, as did his Western sociological counterparts, compared India's caste system to the polymorphism found in the societies of social insects: "It is interesting to note that in East Asia, where we find the organization of social insects developed to perfection, there also has been seen among human associations a minute and even rigid specialization of functions, along with ant- and bee-like societal integrity and cohesiveness. The bionomical resemblances between insect associations and caste-ridden societies, indeed, are amusing" (1926, 228).

The concept of caste was also used in a generalized way to discuss and compare occupational specialization. Emory Bogardus used the term "caste" to describe human occupational specialization just as Wheeler used it to describe occupational differentiation in insects. Although both sociologists and entomologists generally pointed out the differences in the formation of caste for humans and insects, the overarching principle of caste was the same. The entomologist Wheeler noted that caste for humans in modern society as described by Durkheim becomes expressed through professional divisions, rather than being in place from birth as for insects (or positions in traditional society). Despite the differences between societies of humans or insects, both depended on the principle of a rigid specialized division of labor that included castes. "This whole subject has been so thoroughly studied by Durkheim (1922) that it need not be considered here" (Wheeler 1928a, 309). Wheeler adopted Durkheim's ideas about division of labor and caste because he thought they explained social cohesion in a comprehensive and universal way. Following from Durkheim and then Wheeler, caste became an important aspect of social cohesion for both fields.

Mukerjee used both Durkheim and Wheeler in his discussions of caste and occupational specialization. He vacillated on whether occupational specialization was as rigid for humans as it was for insects, yet he provided examples of physical atrophy in both insects and humans that emerged as a result of a high degree of specialization. He believed that the termites with their male and female worker caste were more similar to human society than the ant colony's female-only worker caste system. "Here the sexes are of equal social importance, each caste comprising individuals of both sexes, whereas in ants the workers and soldiers consist of female individuals only. Workers and soldiers among the termites seem to be suppressed males as well as females. Thus we find a more marked resemblance with the ecological pattern of human society" (1942, 35).

In Bogardus's use of the caste concept, the rigidity of occupational specialization in human society was not physical but social; and this is not of minor consequence. Bogardus reflects a very stratified social system with clear indications of hierarchy with his use of the terms "higher" and "lower." In his interpretation of the social system, the social division of labor, with its specialized occupations, produced actual castes.

> The caste arises from identity of profession; it is the most compact of all social organizations. After a person has become established in a profession he has become a member of an existing caste and is under its *esprit de corps*. Consider how difficult it is for a man to change from one recognized profession to another line of activity and what contumely is heaped upon the clergyman who changes to the insurance business, upon the lawyer who shifts to bricklaying, upon the teacher who becomes a dairyman. It is disgraceful to change from a higher to a so-called lower calling, even though a mistake was made in the initial choice of an occupation. It is even a doubtful or questioned procedure for a person who has reached middle life to change from a lower to a so-called higher calling, even though the individual has been converted to an entirely new view of life. Nevertheless, this inelasticity in public opinion is on the whole justifiable, despite the fact that in the broad sense it creates castes. (Bogardus [1918] 1920, 215)

Caste as a concept shared by entomology and sociology served to explain and support a social organization structure with a rigid, specialized division of labor. The concept of caste was also used until the late 1940s as a system of hereditary or environmental divisions that justified racial inequalities in America. For sociology, this term was discarded because of the growing awareness of the social nature of racial segregation and the empirical evidence on the decline in segregation. The use of the term to describe occupational specialization faded as occupational and professional trends moved toward more mobility and less permanency. In the field of entomology, the causes for caste shifted from inherited to environmental factors, along a similar timeline for the change in thinking about human caste systems. More recently, caste and the rigid division of labor idea of social organization have been significantly challenged by Deborah Gordon (1999) and others who maintain that caste is not a major factor in the organization of social insects. However, the term is still widely used in the entomological literature.

Superorganism

The concept of a "superorganism" is a holistic, interlocking description of social systems that has been used to describe both human and social insect societies. The term originated in the nineteenth century and is experiencing a revival in its use to describe human social organization. Although its usage has been significantly reduced, it has never been fully eliminated from the entomological lexicon. The main challenge in retaining this concept comes from evidence of the process of natural selection on the individual rather than the group. However, because the group as a unit of analysis continues to appear to some as a viable model for organizational study, it would be useful to explore the earlier use of the superorganism concept by sociologists and entomologists and the perception of its universality that emerged out of the exchange.

Herbert Spencer developed the term "superorganism" to describe human society and social systems in general in the *Principles of Sociology* in 1898. In the following passage, Giddings elaborates on the significance of the term as it appears in a later publication of Spencer:

> It is hardly necessary to say that the most important endeavour in this direction is contained in Mr. Spencer's system of "Synthetic Philosophy." In that great work the principles of sociology are derived from principles of psychology and of biology. Social development is regarded as super-organic evolution. It is a process in which all the organic and psychic phenomena of human life are combined in larger forms of intricate yet orderly complexity. . . .

> In Mr. Spencer's view, society is an organism, not in mere fanciful analogy, as in the "Leviathan" of Hobbes, but really; and not morally only, but physiologically as well, because, in its constitution, there is a division of labour that extends beyond individuals to groups and organizations of individuals. (1896, 8)

Giddings makes a point to emphasize that this idea about society is "derived from principles of psychology and of biology." However, Spencer, by connecting the process of natural and social organization, had made it clear that this was more than a derived analogy; he felt it was an identification of general principles. Throughout his work, Spencer did not rely only on biological evidence but interchanged this with cross-cultural social comparisons. It was not because this concept was derived from biology that it was accepted in entomology but rather because Spencer proposed the concept as an overarching organic view of systems, which was then easily transferable to entomology. Descriptions of the organization of social insects began to include this concept in accounts of colony structure, and the entomologist most noted for his adoption of this term was William Morton Wheeler. From lectures he gave in Paris in 1925 and later published in *The Social Insects: Their Origin and Evolution,* Wheeler describes the colony using the term "super-organism": "We have seen that the insect colony or society may be regarded as a super-organism and hence as a living whole bent on preserving its moving equilibrium and its integrity" (1928a, 230).

Wheeler, following Spencer, identified the highly specialized division of labor in ant colonies as a superorganism to be compared with the less-specialized insect organizations. Worster notes that Wheeler was influenced by Spencer and applied the organismic concept to both insect societies and human societies as a general term to describe levels of integration in his ecological perspective: "Apparently Wheeler, like Frederic Clements, first came upon this organismic idea through reading Herbert Spencer. And like that polysyllabic genius of Victorian England, Wheeler began stacking up hierarchies of organic wholes. First, at the subcellular level, were 'biophores'; then cells; then the individual organism as a whole. At the next level came the various social groups, from the ant colony to the human family to the nation. And finally, embracing all animate existence, was the greater ecological order" (1985, 321).

Wheeler's application of the idea of the superorganism to social insect colonies was explored later by the entomologist Alfred E. Emerson (Collias 1991, 626). Specifically for Emerson, the concept of a superorganism served as a way to compare organizational similarities between human and insect societies (Mitman 1988, 118–19). It also allowed Emerson to assert his own philosophy of ethics, which emphasized cooperation over competition. Despite criticisms that the superorganism implied a reduction in individual freedom and led to fascism, Emerson firmly believed that the superorganism provided a balance between individual freedoms and social control that led to social stability (Mitman 1988, 157–60).

The literature on the interchange between biology and sociology tends to emphasize the influence of biology in the formation of the discipline of sociology. However, it is usually assumed that in the long run biological analogies had a deleterious effect when applied to the study of human societies, causing sociology to abandon the connection to biology. Nonetheless, social science is still seen by some to be indebted to biology for its objective methods. Far from this picture of neutral biologists being emulated by social scientists, what emerges from a look at interactions between entomologists and sociologists is their shared commitment to social issues and the hope that these could be addressed through the application of universal principles. Natural scientists were not only influenced by their overall social context but also even at times actively engaged in interactions with social scientists and/or their work. The exchange of ideas between entomology and sociology gave each discipline important "evidence" in its quest to outline ideas of sociality and social organization. Both the evidence from the natural world and the social world helped to create shared concepts and theories and reinforced the overarching paradigms of evolution and ecology.

Sociology as a discipline moved away from some of the obvious trappings of its earlier connection with entomology. However, for entomology, concepts that "socialized" insect behavior remained. A rejection of anthropomorphism surfaced briefly but appears to have been replaced with what Kennedy (1992) calls the "new anthropomorphism," a type of "mock analogy" deemed acceptable and used as a convenient tool for description. Sociology turned away from the terms "caste" and "superorganism" in the 1940s, while entomology continues to use them. With the introduction of Hamilton's inclusive fitness theory, entomology moved away for a time from the concept of a superorganism. However, as later challenges to the theory of inclusive fitness mounted, the superorganism concept began to be reintroduced (Wilson and Sober 1989). Examining specific concepts such as division of labor, caste, and superorganism better illustrates the effect of interactions between sociology and entomology. Entomology, as opposed to sociology, retains some terms such as "caste" and holds onto the concepts of a rigid division of labor and of a superorganism. For both disciplines, however, reemerging theories and concepts—especially concerning organizational structure—are beginning to be shared once again. Social organizational theory has been exploring ideas inspired by computer networking, as self-organizing models replace the more conventional view of a hierarchical social structure. New ways of looking at the organization of social insects have led to alternative biological models influenced by this overall shift in organizational theory. As these models become available, they are taken up as evidence that reinforces new theories and concepts for both entomology and sociology. A mutual interaction of ideas between sociology and entomology is perhaps emerging once again. A closer look at these interactions will reveal a complex exchange of ideas and influence just as occurred between these disciplines in the nineteenth and early twentieth centuries.

5

Despite the Differences
INSECT SOCIALITY AND
COMPARATIVE METHOD

The use of analogies between insect and human societies has been dismissed as a relic of the past, of simple anthropomorphism that at least science has moved beyond. While some consider it unnecessary to explore these antiquated analogies because of the progress made by science in this area, terms from these analogies are still utilized in scientific discussion, and their basic premises are embedded in social theory. In addition, there has been a recent push for the acceptance of anthropomorphic language and analogies, with claims that they are either natural, universal, important in challenging the idea of boundaries between nonhuman and human animals, or necessary in teaching biological concepts (Arluke and Sanders 1996; Beck 1996; Irvine 2004; Sanders 2003; Zohar and Ginossar 1998).

The terms outlined in chapter 1 are used in social/ biological analogies. Because terms and concepts are important keys to the structuring of theoretical knowledge and practice, these analogies become "evidence" in theoretical paradigms and models for social organization and roles. These analogies need to be examined for their meanings, their directionality, and their limitations involving the theories they support concerning both insect and human societies. As Lakoff and Johnson suggest, the first step is developing an awareness of the embedded quality of metaphoric language and how important it is in forming the conceptual structure around which we organize our lives. Because the conceptual system is behind our communication concerning cognitive and behavioral processes, language is the key to identifying the properties of that conceptual system (1980, 3).

Analogies comparing social insects to humans were used extensively through the nineteenth and early twentieth centuries and are currently experiencing a revival. Some of the same hierarchical language is used in the current scientific discourse; therefore, these analogies still play a legitimizing role in reinforcing structures. As Lakoff and Johnson point out, even words like "high" and "low" can imply and reinforce a hierarchical system because of the meanings attributed to them. Newer terms such as "self-organizing" and "hive mind"— terms in common use today—are contributing to new ways of conceptualizing social systems. The naturalizing of social systems through the use of metaphoric language or the misinterpretation of natural systems through social terms still very much need to be explored.

While the use of metaphor and analogy in everyday language is acknowledged, it is often imagined that science uses neutral language. However, scientific discourse routinely features analogy and metaphor. Hesse (1966) marks their usage from Aristotle onward in the development of scientific models and theory building. Scientific analogies are more than simple comparisons between a familiar and unfamiliar domain; they also assume a connection to an overarching natural law or a universal correspondence, according to Hesse. Citing Black's (1962) "interaction view" of metaphor, she notes that each domain loses some of its property through an analogy and becomes understood to have a "postmetaphoric" meaning (163): "Men are seen to be more like wolves after [a] wolf metaphor is used, and wolves seem to be more human" (163). From this merged identity, comparisons can become universals.

Analogy and metaphor are often discussed as if they were interchangeable concepts because they are so closely intertwined. Klamer and Leonard explain the relationship of metaphor to analogy within scientific usage: "Analogy is an expanded metaphor; more precisely, analogy is sustained and systematically elaborated metaphor. Accordingly, in a scientific context, a metaphor becomes heuristic when it stimulates the construction of an analogical system. The mere coinage of a metaphor such as "human capital" does not make science" (Klamer and Leonard 1994, 35).

The terms used to describe the roles of social insects tend to be metaphors that are then elaborated on within analogies. For instance, a typical sentence might state, "A nonreproductive bee is a factory worker"; the same notion might also be expressed with a simile, "A nonreproductive bee is like a factory worker." Both make a comparison, but a metaphor is a more direct transfer of qualities. An analogy might then employ the metaphor in a comparison: "A manager oversees the making of widgets, just as the queen directs the workers to produce honey in the bee factory." An analogy can take into account differences as well as similarities; but it also can imply that there are even more points of connection and that the connections are real, not just a figure of speech or poetic license. The analogous statement above can be expanded to describe connections between specific tasks, human and insect "cities," or any number of things deriving from this seemingly simple initial comparison. As Lewontin (1991, 95–96) suggests: "What happens is that human categories are laid on animals by analogy, partly as a matter of convenience of language, and then these traits are 'discovered' in animals and laid back on humans as if they had a common origin." He notes that analogous similarities do not always indicate actual similarities in either their biological or social purpose.

Others have also claimed that the use of social terms to describe natural phenomena can cause bias toward and distort perception of the object of study (Hull 1992; Kennedy 1992; Lewontin 1991; Margulis 1998; Owen 1990); that the issue with metaphors or analogies in science is that they can become literal, be incorrect, or both (Cherry 1988; Duit 1991; Knorr-Cetina 1981). Lynn Margulis offers harsh criticism for the continued use of terms that do not adequately reflect what they seek to describe: "Language can confuse and deceive. These antiquated terms—'blue-green algae,' 'protozoa,' 'higher animals,' 'lower plants,' and many others—remain in use despite their penchant to propagate biological malaise and ignorance" (1998, 55).

I share the position that metaphoric terminology and analogies used in describing human and nonhuman social behavior can lead to misunderstandings, bias, and limitations in research and theoretical perspectives. A classic example exists in the paradigm shift that occurred in primate studies (Asquith 1996; de Waal 1989; Haraway 1989; Marks 2002; Strum and Fedigan 2000). There had been many studies of primates who were described as territorial, male-dominated, violent, and competitive; this was interpreted as offering a definitive and natural explanation for this type of human behavior as well. At some point, primatologists—most notably women primatologists—brought forth the bias in this analogy between humans and primates and especially in the choice of studying certain primates that confirmed aggressive and competitive behavior to the exclusion of studying other primates that exhibited more cooperative behavior.

In the field of entomology, anthropomorphic terms continue to be used to describe social insects and their behavior. Social insects are described as workers, queens, soldiers, farmers, slaves, and army ants. Their organization is described as a monarchy, a democracy, a caste system, and as maintaining a rigid division of labor and hierarchy. Their soldiers create bivouacs and launch raids and maneuvers; potential queens take marriage flights; sisters exhibit altruism; farmers harvest crops; workers labor for those higher up in the hierarchy. Because of their *social* designation and the history of comparison to human social organization, these terms are striking. These terms are still applied to humans and human organizational structures as well. The social analogy between insects and humans is embedded in the language itself. These terms allow for a "natural reflection" of human social structure, continuing the purpose of legitimation as needed.

DEBATING THE DIFFERENCES

The use of analogies between insect and human societies and the arguments for or against using them were frequent in the literature of the late nineteenth and early twentieth centuries. This literature included outlining the similarities and differences between human societies and insect societies, following the general pattern of evolutionary schemas, so that certain differences in human society were seen as a distinguishing step that identified humans as "higher" or more evolved than the social insects. While that one step was not always agreed upon, in most arguments insect societies were used as a straw man that helped establish the claim of uniqueness for human societies.

These distinctions were especially important for sociologists to make as they broke away from associations with biology. To claim something unique about human society as opposed to insect societies gave sociologists a distinct "territory" to analyze. However, just what that territory would be was not always easy to determine. The distinctions were problematic, considering that what some authors considered to be unique to humans was seen by others to be a quality shared with social insects. The question of what makes humans unique is not at all resolved even today as we note that DNA samples place us firmly in relation to primates, and the lines between human minds, societies, and machines becomes blurred (Haraway 1991; Kelly 1994; Marks 2002). The scientific community's use of the new anthropomorphism also challenges the divide between human animals and other animals.

There is a renewed discussion of anthropomorphism and of the difference between humans and nonhumans (Alger and Alger 1997; Arluke and Sanders 1996; Crist 1999; Irvine 2004; Kennedy 1992; Ritvo 1999; Rumbaugh 1999; Sanders 2003). Some authors argue against or defend the use of anthropomorphic analogies, while others strive to break down boundaries between what they see as falsely constructed differences betweens humans and nonhuman animals. As Kennedy (1992) notes: "We are witnessing a new swing of the theoretical pendulum, now back towards anthropomorphism" (5). This neoanthropomorphism, as Kennedy terms it, even becomes a part of the neobehaviorist's way of describing animal behavior as they cross the line from "mock anthropomorphism" into genuine, unconscious anthropomorphism. Analogies that are anthropomorphic can be confusing for readers and, Kennedy claims, for scientists as well. Kennedy therefore warns about an unrestrained use of anthropomorphic analogies but also believes that their use may be a hardwired cognitive tool that can never be fully eliminated. Crist (1999) sees no reason to attempt to eliminate anthropomorphism and instead advocates for it, viewing the case against anthropomorphism as fostered by a naïve belief in neutral language and an artificial separation between nonhuman animals and humans. Alger and Alger (1997), Arluke and Sanders (1996), Irvine (2004), and Sanders (2003) also question the theoretical separation of nonhuman animals from humans. Exploring the world of companion animals, they point to a shared interaction between nonhuman animals and their human companions as evidence of the common ground between species.

Some of these authors evoke the earlier discourse on the topic to defend or support their own position. Rumbaugh (1999) presents Darwin's writing on the continuity between animals and humans in his argument for rejecting the Cartesian mechanical model of animals. Irvine (2004) refers to Darwin's work on emotions and animals and challenges Mead's definition of a selfhood centered on language. Alger and Alger (1997) contend that Mead "drew a hard line" dividing humans and nonhuman animals. Their work on companion animals intends to go "beyond Mead" in presenting evidence of symbolic interaction between domesticated cats and humans. Expanding the notion of self and social roles from established sociological insights may shift the emphasis of research just as debates in the nineteenth and early twentieth centuries once did.

Much of the new research on this topic revolves around companion animals or primates. Added insights can be gained from a look at invertebrates, in particular social insects and the earlier discussion that took place over anthropomorphic analogies and comparisons with these species. These discussions took place in the larger context of evolutionary theory and discipline formation. Evolutionary debates at that time utilized comparative examples between animals and humans, and frequently social insect societies were the model used.

SOCIAL INSECT SOCIETIES AS A COMPARATIVE SOURCE

While in the late twentieth and twenty-first centuries comparative debates predominantly involve comparing primates to humans, during the nineteenth and early twentieth centuries this was not the case. Despite Darwin's revelation of the evolutionary relatedness of primates and humans, by and large no systematic scientific comparisons were being done. One reason for this was that primate studies were not usual in the nineteenth and early twentieth centuries so that data for comparison were not readily available. The studies that were being done were largely on captive animals in zoos. Not until the 1950s did field studies really began to present what was then considered significant data about primate society that could be used as a basis for comparisons to human society (Ardrey 1969: Burton 1994). Robert Ardrey (1969) claims, in the introduction to Eugène Marais' book *The Soul of the Ape,* written in 1927, that earlier field studies such as those conducted by Marais and Clarence Ray Carpenter were largely ignored; therefore Ardrey identifies the watershed for primate studies as Sherwood Wasburn's and Irven De Vore's 1961 paper "The Social Life of the Baboon." Robert Yerkes, another pioneer in primate studies, wrote of a need for a primate laboratory as early as 1916 in an article for *Science.* However, the Yale Laboratories for Primate Biology did not open until 1930 (www.yerkes.emory.edu). While the cumulative work accomplished by Marais, Yerkes, Carpenter, and others certainly caused an eventual shift to the use of primates as the ultimate comparative behavioral model for humans, the use of other models predominated before this occurred.

Other frequent models of comparison in the nineteenth century were the body and the cell. Comparisons to social insects complemented the model of the cell and the body because the social insect colony was seen as a "superorganism" and therefore served as an example for the organicist theory as did the cell and the functions of the body. As opposed to the cell and the body, specific advantages to the use of a species so very different from humans would be that the similarities found across these vast differences seemed to suggest a common natural law. It rose above the solipsism of using the human body to illustrate how human society works. Just as in the case of primates currently, it offered something "outside" the human form to compare behavior and social structures.

Social insects also had an evolutionary advantage—the established fact that they had evolved and maintained societies well before humankind. Remy Chauvin describes the credibility of social insects as a source of comparison in his contemporary book *Animal Societies: From the Bee to the Gorilla*: "[W]ith vertebrates, even primates, we are in a world just before the Stone Age, whilst in the case of the social insects we are studying a *civilization* infinitely older than that of man. It must be understood that we are talking of an *insect civilization*, still utterly different in kind from our own. But if by 'civilization' we mean the development of complicated social organization, the carrying out of works in common, the methodical care of the young by the community, and the division of labour among specialized groups, then undoubtedly we have the right to use this word" ([1963] 1968, 10).

DEBATES ON THE COMPARATIVE METHOD

In the late nineteenth and early twentieth centuries, sociology and entomology had to address the issues framed by comparative psychology, as did most other emerging disciplines and subfields of the time. The comparative method made influential by Herbert Spencer, George Romanes, Charles Darwin, and others was being challenged but still remained the primary scientific method (Robinson 1977). Sociologists and entomologists weighed in on the debates involving comparative psychology and the comparative method. Charles Ellwood, G. H. Mead, Emory S. Bogardus, Franklin Giddings, William McDougall, Auguste Forel, Sir John Lubbock, William Morton Wheeler, and Alfred Emerson were just a few of the prominent figures in these debates who addressed the issues and sometimes each other's views on the topic as well.

In general, even when comparisons to animals were not the direction in which some sociologists wanted to go, there seemed to have been some acquiescence to the notion that analogies could be helpful. Ironically, even within the debates about whether the analogies were adequate, analogies about social insects were employed to argue that they should not be used. Max Weber, addressing the limitations of comparisons of animals to humans, concedes their possible future importance for issues pertinent to sociology: "[I]n the field of animal psychology, human analogies are and must be continually employed. The most that can be hoped for is, then, that these biological analogies may some day be useful in suggesting significant problems. For instance they may throw light on the question of the relative role in the early stages of human social differentiation of mechanical and instinctive factors, as compared with that of the factors which are accessible to subjective interpretation generally, and more particularly to the role of consciously rational action" (1978, 17). The general discourse of the time was informed by evolutionary theory, which supported the comparative method; therefore it was necessary for social scientists to support or refute the premises of these larger debates while addressing the methods used for these claims. As disciplinary boundaries were not clearly drawn at that time, it would be difficult to avoid the primary discourse on matters relating to the nature of the self and society.

Instinct

There were many references within the dominant discourse to the extent of instinct-driven behavior in human or insect societies. The sociologist Charles Ellwood (1901, 729–30) believed that the instincts of insects provided a baseline of development on which one could model human organization:

[I]t is usually recognized that the organization which colonies of these insects exhibit is an outcome of certain habits of cooperation which have become *innate* in the species through a process of natural selection in the course of a long period of evolution. In other words, the societies formed by ants, bees, and wasps are organized upon the basis of instinct. Now, if instinct plays such a rôle in the organization of sub-human societies, and if human societies are admittedly genetically related to these, is it not probable that instinctive impulses have much to do with the organization of human society; and not simply one instinctive impulse, the tendency to imitate, but many?

Even if humans evolve to higher states than "sub-human societies," their social organization always traces back to its instinctual base. The comparative connection between social insects and humans appears to confirm that for Ellwood.

Those who would make a comparison between humans and social insects in regard to instinct had to first argue that instinct existed. They then had to argue that the behavior of insects in particular was driven by instinct rather than by any form of intelligence. Franklin Giddings referred to the frequent discussion of instinct and comparisons to animals, including social insects: "A swarm of bees, a hill of ants, may be instinctive, but, of course, you know that a question has been raised of late as to whether there is any such thing as instinct. But let us concede for the moment that there is" (1932, 363). Having established a hypothetical condition for instinct, he continued, noting that social insects proved the most challenging case:

The question is often raised whether the so-called marvelous instincts of the bees and the wasps, are strictly speaking, mental. . . . When bees or wasps of the same swarm are friendly with one another and are hostile to those of other swarms, it is almost certain that it is not a mental fact but a purely chemical lot of changes or tropisms acting through the nerves through the sense of smell. There is a vast field for study here which has not yet been worked over. It will probably result in showing to us that we are dealing with phenomena that in the last analysis are physical and chemical and that only as we understand the physical and chemical facts we will be able to understand a lot of the mysteries of human conduct. (1932, 18–19)

Giddings maintained that social insect behavior was instinctual and, along the lines of Ellwood, thought there was still an evolutionary connection that allowed for useful comparative data to be derived from this example. Instinct was therefore viewed as important to the human experience and was evolutionarily illustrated through social insects.

Durkheim described instincts as "unconscious," "perfect," and "immutable." He used animals as examples because he felt that instinct was more important for animals than humans. Animals are "little more than the playing out of a series of instincts" (Gross and Jones 2004, 148). Durkheim believed that for humans, instinct plays a decreasing role as they become adults. However, he also claimed that instinct is psychological—not physiological—and that therefore insects (and children) are not mere automata. So, although "[i]nstincts are immutable—the same today as they've always been (bees, for example, make their honey as they always have)," he also notes that "this immutability isn't absolute, for the influence of man or environment can change instinct" (Gross and Jones 2004, 148–49).

Emory Bogardus appeared to use the comparison to social insects in a more pluralistic fashion that included both insects and humans, placing them at various points on a scale of instinctive to purposive: "Permanent groups vary from purely instinctive to socially purposive. The best illustration of purely instinctive grouping is found among animals, e.g., insect societies. The primitive horde and the family are less instinctive than an insect society" ([1918] 1920, 214). This would seem to indicate that Bogardus differentiates social insect societies from human societies based on some evolutionary criteria, with only insect societies defined fully by instinct. The difference is seen as one of degree since the human groupings are seen as "less instinctive." However, he goes on to say: "The modern family including courtship is often instinctive, although showing a few signs of conscious purpose that are worthy of these institutions. The modern state is largely instinctive. . . . Economic organizations, such as corporations and labor unions, are distinctly purposive. Educational associations are strikingly telic" (214). Bogardus sees certain human institutions, such as the state, as instinctive and therefore as inviting comparison to insect societies.

Lester Ward also wrote about similarities of state structure in the societies of social insects and humans, noting that social insects derived this structure from instinct. He believed that "certain forms of sociability appear among creatures to which intelligence cannot be imputed, not merely among many of the higher mammals and other vertebrates, but notably among insects. Here instinct seems to have brought about the same general economic system that has resulted in part at least from rational calculation in man" (1895, 145). Whether from instinct, as in the case of social insects, or instinct enhanced by reason, as in the human case, similar social and economic institutions were created in human and insect societies, which did seem to create a foundation for comparing these social structures.

Entomologists were also interested in the role of instinct in creating similar structures and behaviors of social insects and humans. Alfred Emerson (1942, 174) believed that one of these shared characteristics was an ethical system: "With the shift from the germinal and physiological factors which obviously coordinate the organism and social supraorganism in the biological world to the social heredity through learned symbols in the human world, we find that we have arrived at an ethical system. Ethical patterns as coordinating factors in human society have their parallel in the physiological integrating factors of the organism and social insect supraorganism." Just as some sociologists pointed out, the structural or behavioral similarities were worth comparing, despite the level of instinct that led to the particular structure or behavior.

Other entomologists began to lean more toward some semblance of shared reason in the development of these structures or behaviors, which would further justify the use of comparisons. In an interesting "confession," Auguste Forel claimed that he overstated the case of instinct in an earlier work so as not to be criticized for anthropomorphism:

> Being afraid of the prejudices of science, which at that time was in re-action against the childish anthropomorphism of the past and therefore regarded all animals, especially all insects, as little more than reflex machines, I felt obliged in my *Les Fourmis de la Suisse* to confine myself exclusively to well-authenticated facts.

> In many respects, however, these facts still remain a mystery to me. I felt that before the ant-mind could be understood, it would be essential to go deeper into the study of the sensations of ants and of insects in general, and I began to observe them zealously by means of some new experiments. (Forel 1928, 26)

Forel subsequently attributed more than instinct to social insects. He believed that social insects had a range of emotion, including some feelings that were instinctual and some situational: "[F]eelings of sympathy, antipathy, and anger among ants may be intensified by repetition and by the corresponding activities, just as in other animals and man. The great social sense of duty is instinctive in ants, though they exhibit great individual, temporary, and occasional deviations, which betray a certain amount of plasticity" (1904, 33).

One idea that emerged in this area was that the particular instinct displayed by social insects was actually an evolved mechanism that allowed for greater social integration. In this regard, instinct had a different meaning and was seen as not a lesser trait indicative of lower intelligence but rather a way to free up that intelligence for greater things. In this, social insects were portrayed as a model for the future evolution of sociality in humans. An example of this can be seen in Eimer's *Organic Evolution* (1890, 424–25):

> Since these actions are performed automatically, the animals are left free to exert themselves in other ways: they have time, opportunity, and energy to apply their mental activity in other directions; and we can even see in the development of such a state of things the ideal condition of the state, since the individual performs his part for the good of the whole more and more mechanically, and as a matter of course, and is no longer compelled to consider on each occasion whether he must do it, and how.

Consciousness

Closely related to the question of instinct was the existence of consciousness and reason in social insects as compared to humans. If consciousness was to be assumed, then the question became one of determining if an individual consciousness or a collective consciousness characterized human or insect societies. Therefore comparisons to animals often centered around whether humans differed from herds or colonies that were more often seen as driven by a collective consciousness. William McDougall addressed this in his book *The Group Mind* (1920, 92):

> The group spirit, involving knowledge of the group as such, some idea of the group, and some sentiment of devotion or attachment to the group, is then the essential condition of all developed collective life, and of all effective collective action; but it is by no means confined to highly developed human associations of a voluntary kind. Whether the group spirit is possessed in any degree by animal societies is a very difficult question. We certainly do not need to postulate it in order to account for the existence of more or less enduring associations of animals; just as we do not need to postulate it to account for the coming together of any fortuitous human mob. Even in such animal societies as those of the ants and bees, its presence, though often asserted, seems to be highly questionable.

Although McDougall believed that "group spirit" might not exist in social insects, he also realized that it is not attributable to humans groups alone. He seemed to indicate that while it may have been probable for social insects, the question of instinct remained. He did not appear to attribute group consciousness to a human mob either and did not belabor whether it could be. Yet, McDougall did elaborate on the case of social insects because they were a part of the dominant analogies and were a topic in the work that was establishing disciplinary boundaries:

> When we observe the division of labour that characterizes the hive, how some bees ventilate, some build the comb, some feed the larvae and so on; and especially when we hear that the departure of a swarm from the hive is preceded by the explorations of a small number which seek a suitable place for the new home of the swarm and then guide it to the chosen spot, it seems difficult to deny that some idea of the community and its needs is present to the minds of its members. But we know so little as yet

of the limits of purely instinctive behaviour (and by that I mean immediate reactions upon sense-perceptions determined by the innate constitution) that it would be rash to make any such inference. (1920, 92–93)

McDougall made a distinction between what he called "group spirit" and collective behavior in general. Although McDougall remained unsure of the existence of group spirit among social insects, he seemed to feel that it was a verifiable aspect of human society. He disagreed with the unqualified assignment of a "collective consciousness," as espoused by Espinas (1878) in his example of bees and ants, as well as by Schäffle and other organicists who used biological analogy to indicate a direct correspondence rather than a similarity.

Regarding the collective nature of consciousness, Eugène Marais, in his research on termites, would not have agreed with McDougall about a lack of group spirit in social insects; instead, he found termites to be one of the best examples of this. In order to "understand even a little about the behaviour of the termite," Marais said one must know about the "group psyche" (1937, 59). He compared this to the way the human body functions and believed that, "After the composite physical body of a highly developed animal like man, there is no better example of the functioning of a group soul than the termitary" (1937, 60). However, Marais pointed out the mistake made in the comparisons of the consciousness of humans and social insects. "We humans use our own consciousness as a criterion for classification. Eventually we discover that this consciousness can never be a criterion for psychological processes different from our own" (1937, 74). According to Marais, the human psyche should not be the only standard for understanding consciousness.

Durkheim had his doubts about comparison as well but for different reasons. Unlike Marais, he did not attribute any consciousness to insects, believing that, "if instinct were conscious, then animals would possess a sense of foresight infinitely more developed than that of human beings. To believe that bees consciously build the combs destined to receive their honey, we must believe that they understand geometry" (Gross and Jones 2004, 88). On the other hand, Durkheim maintained that instincts were adaptable and could be changed by external factors. It is just this adaptability claimed by Durkheim that spurred other sociologists and entomologists to attribute consciousness to insects. Charles Ellwood represents this line of thinking, arguing that, "we cannot well deny to these creatures some degree of mental life, for they are known to show, both as individuals and as groups, considerable power of adaptation in the presence of danger" (Ellwood 1901, 729). Evidence of adaptation became the enhancement that elevated social insects to the level of having consciousness. The question still remained, how much did they consciously reason as compared to humans?

For the entomologist Sir John Lubbock, the consciousness of social insects was not a difference in kind, as Giddings would later say, but only of degree:

> In face of such facts as these, it is impossible not to ask ourselves how far are ants mere exquisite automatons; how far are they conscious beings? When we see an ant-hill, tenanted by thousands of industrious inhabitants, excavating chambers, forming tunnels, making roads, guarding their home, gathering food, feeding the young, tending their domestic animals,—each one fulfilling its duties industriously, and without confusion,—it is difficult altogether to deny to them the gift of reason; and the preceding observations tend to confirm the opinion that their mental powers differ from those of men, not so much in kind as in degree. (1882, 181)

And Charles Darwin, whose work influenced both sociologists and entomologists, exalted the mental abilities of social insects: "It is certain that there may be extraordinary mental activity with an extremely small absolute mass of nervous matter: thus the wonderfully diversified instincts, mental powers, and affections of ants are notorious, yet their cerebral ganglia are not so large as the quarter of a small pin's head. Under this point of view, the brain of an ant is one of the most marvellous atoms of matter in the world, perhaps more so than the brain of man" (1874, 54). Darwin notes both instincts and mental powers (as well as affect) in ants, praising their ability to have both instinct and conscious reasoning in such physically small-sized brains. Humans come up short in comparison when matched on relative scale.

If consciousness was to be considered, then the question of language had to also be addressed by sociologists and entomologists. Although language seems to be one of the main differences between social insects and humans for Mead and others, Mead himself notes that it may be premature to claim this skill as a characteristic only of humans:

> All communication, all conversations of gestures, among the lower animals, and even among the members of the more highly developed insect societies, is presumably unconscious. Hence, it is only in human society—only within the peculiarly complex context of social relations and interactions which the human central nervous system makes physiologically possible—that minds arise or can arise; and thus also human beings are evidently the only biological organisms which are or can be self-conscious or possessed of selves. It is a distinction which still has to be made with reservations, because it may be that there will be some way of discovering in the future a language among the ants and bees. (1934, 234–35)

Forel (1904) made the claim that nonverbal communication in humans might be more accurately interpreted than verbal language because words have a greater chance of being misinterpreted. Therefore he found that, "the attitudes and behavior of animals have for us the value of a 'language,' the psychological importance of which must not be underestimated" (1904, 6). He felt that the main difference between the languages of other animals as compared to human language is that they did not contain abstractions; therefore Forel (1904, 21) maintained that this language could not be compared with human language: "That ants, bees, and wasps are able to exchange communications that are understood, and that they do not merely titillate one another with their antennae as Bethe maintains, has been demonstrated in so many hundred instances, that it is unnecessary to waste many words on this subject. . . . But, of course, this is not language in the human sense!" Lubbock (1882, 181) claimed that there was empirical evidence of some type of language abilities in social insects, stating, "These experiments certainly seem to indicate the possession by ants of something approaching to language."

Maeterlinck (1901, 178) proposed that it was important to know if social insects had a consciousness leading to reason and a type of language because this knowledge might contribute to an understanding of human consciousness.

> Were we sole possessors of the particle of matter that, when maintained in a special condition of flower or incandescence, we term the intellect, we should to some extent be entitled to look on ourselves as privileged beings, and to imagine that in us nature achieved some kind of aim; but here we discover, in the hymenoptera, an entire category of beings in whom a more or less identical aim is achieved. And this fact, though it decide nothing perhaps, still holds an honourable place in the mass of tiny facts that help to throw light on our position in this world.

The Role of the Individual and the Group

For Mead, the role of the individual in society was intertwined with society itself, as a certain type of society produces a particular individual. Mead denied that differentiation alone created a complete similarity between social insect society and human society. He considered this to be an "utterly illogical type of analysis which deals with the human individual as if he were physiologically differentiated, simply because one can find a differentiation of individuals in the human society which can be compared with the differentiation in a nest of ants" (1934, 244). Mead believed that human society went beyond instinctual collectivity; in human society the individual's role is understood in relation to others in a conscious way. If differentiation is formed in a society by language, as in human society, it produces conscious awareness that in turn creates a different individual from the one created by physiological differentiation. "In man the functional differentiation through language gives an entirely different principle of organization which produces not only a different type of individual but also a different society" (1934, 244).

Although Mead claims that human society differs from animal society, he also uses a comparison to social insects to put forth one of his strongest theses about the individual and society: that society is prior to the individual and that individuals are created out of societal interaction. To set up the analogy, he first describes a bee colony, and when describing physiological differentiation, he notes that the extent of this differentiation could emerge only from a society:

> There could not be the peculiar development found in the beehive except in a bee community. . . . In such organizations as the beehive there arise the conditions under which, due to the abundance of food, the forms carry over from one generation to another. Under those conditions a complex social development is possible, but dependent still upon physiological differentiation. We have no evidence of the accruing of an experience which is passed on by means of communication from one generation to another. Nevertheless, under those conditions of surplus food this physiological development flowers out in an astonishing fashion. Such a differentiation as this could only take place in a community. The queen bee and the fighter among the ants could only arise out of an insect society. One could not bring together these different individuals and constitute an insect society; there has to be an insect society first in order that these individuals might arise. (1934, 232–33)

Mead then follows this passage with a similar one about human individuals in a society. Mead finds it useful to use the analogy of the differentiation of social insects to put forth his idea on the relation of individuals to society, showing his willingness to trade on the naturalizing example commonly found in the discourse of the time. Drawing a comparison to the societies of social insects strengthened his argument about human societies. The analogy draws on a more familiar individual differentiation emerging from an insect society to what Mead then proposes is only a more subtle manifestation of differentiation emerging from human society:

> In the human community we might not seem to have such disparate intelligences of separate individuals and the development of the individuals out of the social matrix, such as is responsible for the development of insects. . . . It is because of such considerations that a theory has developed that human societies have arisen out

of individuals, not individuals out of society. Thus, the contract theory of society assumes that the individuals are first all there as intelligent individuals, as selves, and that these individuals get together and form society. On this view societies have arisen like business corporations, by the deliberate coming-together of a group of investors, who elect their officers and constitute themselves a society. The individuals come first and the societies arise out of the mastery of certain individuals. The theory is an old one and in some of its phases is still current. If, however, the position to which I have been referring is a correct one, if the individual reaches his self only through communication with others, only through the elaboration of social processes by means of significant communication, then the self could not antedate the social organism, the latter would have to be there first. (1934, 232–34)

Mead therefore makes a parallel argument for social insects and humans in that the individual emerges from a society. He states that there is no evidence of communication in insect society, but that despite this the complex differentiation of their societies illustrates his idea that the community precedes the individual. Social insects cannot exemplify his condition that the individual becomes a self "only through the elaboration of social processes by means of significant communication." This does not contradict his comparison if it is seen in terms of the *degree* of an individual's relation to society, the difference between the meanings of the term "individual" and "self." The various roles of individuals in a bee colony are not elevated to the level of the role of an individual who becomes a "self" in a human community; "selfhood" becomes a separate matter that applies to human society.

The sociologist Alfred Espinas (1878) was interested in the effect of a crowd on the individual, claiming that behavior changed when an individual was isolated or interacting within a crowd. Using an analogy between the influence of a speaker on a crowd and the behavior of an ant, Espinas attempted to universalize the significance of the collective. He described a multiplication effect of "300 times" the original feeling because of the reaction from a crowd: "[W]hen the orator manages to dominate his emotion and reacts to the crowd, we can see what repercussions of electric shocks must be established between him and his audience and how one and the other are almost instantly transported far beyond their habitual moral realm. The same thing happens in any reunion of feeling beings, no matter what they are; not only does the emotion of a single being communicate to other, but moreover, the bigger the agglomeration is, the more intense the emotion grows. The observations of M. A. Forel are an exemplary verification of it" (1878, 363). To tie the analogy to a natural law, this sociologist used entomological data from Forel on the effect of the group on a solitary ant, adding that, "the ant that would die 10 times when fighting alongside fellow ants, turns out to be very shy when isolated" (1878, 363). He also utilized information from a naturalist's observation of wasps in that "the more numerous they were, the more agitated they were and the faster that emotions transferred to other wasps" (1878, 363).

An issue that appears frequently in the dominant discourse is one of the extreme sacrifice of the individual for the sake of the larger society that seemed to be exhibited by social insect behavior. This is often focused on as an ideal type of behavior, yet at times it is discussed with some caution as well. A parallel was made about the problem in human society of those who were not seen as contributing to the common good. The harsher regulation of the individual in social insect society was seen as a possible prescriptive for human society. The following passage by Eimer (1890) is a good example of this type of comparison:

> The state-organisation of bees and ants has often been praised as the ideal of social and political arrangements. But it must be remembered that in these animal communities the individual is unconditionally responsible to the state for all its activity, as in the state's servant, and if it does not fulfil [sic] this responsibility, or if it has served its purpose, it is expelled or destroyed. The queen-bee does not begin to provide for the reproduction of the hive until there are a sufficient number of workers to secure adequate maintenance for two hives, and the female humble-bee or wasp likewise begins her family by providing workers. This is according to reason; while the causes of the social problems in mankind depend in great part on unreasoning conduct with respect to the establishment of families, and for the rest on the burden of idlers in the state. (1890, 275–76)

In the above passage, social insects order their communities "according to reason" as compared to humans. The problems of humankind are viewed as stemming from those not contributing to the state and the unlimited freedom of families to self-regulate. Eimer explicitly proposed that the individual and family unit must be subordinate to the collective good and that social insects could be models in this regard: "Thus our conception leads not only to the complete recognition of the rights of our neighbour, but also to the most complete subordination in family and state. It is the most uncompromising opponent of that confused idea of freedom so injurious to the common good which claims unlimited independence for the individual. It takes in some sense the social life of the bees for its model, in which the work of the individual for the community has become automatic action" (1890, 433–34). The order and reason that emerges from social insect communities, according to Eimer, was enabled by instinct culminating in the "automatic action" of social duties.

POSSIBILITIES OF THE COMPARATIVE METHOD

Max Weber suggested in 1913 that current understanding of the behavior of social insects was not complete at that time and remarked on the potential for comparisons beyond the merely functional roles within a social insect society. He discounted as premature any earlier accounts of insect sociality that went beyond functional analysis. As to what was currently possible, Weber determined that, "a purely functional point of view is often the best that can, at least for the present, be attained, and the investigator must be content with it. Thus it is possible to study the ways in which the species provides for its survival; that is, for nutrition, defence, reproduction, and reconstruction of the social units. As the principal bearers of these functions, differentiated types of individuals can be identified: 'kings,' 'queens,' 'workers,' 'soldiers,' 'drones,' 'propagators,' 'queen's substitutes,' and so on" (1978, 16). Merely identifying functional roles for social insects and basic behavior was not seen as an endpoint for the science of social insects. Therefore, according to Weber, "all serious authorities are naturally fully agreed that the limitation of analysis to the functional level is only a necessity imposed by our present ignorance, which is hoped will only be temporary" (1978, 16). More than once, Weber stressed that the limitations of a strictly functional analysis did not reflect the possibilities for understanding the social life of insects. And although in general Weber felt that defining subjective understanding between animals and humans could be "problematical" at the given time, he also believed that, "in so far as such understanding existed it would be theoretically possible to formulate a sociology of the relations of men to animals, both domestic and wild" (1978, 15).

Similar to Weber, there was an "open door" aspect to Mead's discussion of invertebrates, especially as it focused on language and reason, two major characteristics that were (and still are) seen to be distinctive for humans. Although Mead made distinctions between humans and social insects from the information then available, he did this with "reservations." This same caution can be found in McDougall's, Giddings's, and Ellwood's discussions of the value of comparing humans and animals.

Auguste Forel, who at one point was careful to emphasize the differences, began to feel that the similarities were more important: "When discussing the ant-mind, we must consider that these small animals, on the one hand, differ very widely from ourselves in organisation, but on the other hand, have come, through so-called convergence, to possess in the form of a social commonwealth a peculiar relationship to us" (1904, 1). Citing humans' achievement of an objective psychology through the techniques of analogy, observation, and induction, Forel argued that this likewise meant that a comparative animal psychology might also exist (1904, 5). He further surmised: "It results, therefore, from the unanimous observations of all the connoisseurs that sensation, perception, and association, inferences, memory and habit follow in the social insects on the whole the same fundamental laws as in the vertebrates and ourselves" (1904, 20–21). Forel believed that "notwithstanding all the differences," social insects and humans are still tied to the same fundamental laws of the senses.

Emerson (1942, 164) proposed that research comparing insects and humans could take place because of shared social behavior: "The similarities of human, ant and termite societies, therefore, are not due to a common heredity, but does this mean that the common social behavior is not worthy of scientific investigation? Surely we are interested in the causative factors which have directed even dissimilar types toward similar social characteristics." He encouraged collaboration among sociology and entomology in pursuing these comparisons in order to better understand fundamental principles of organization.

Even more, the idea that social insect organization and behavior could serve as a model was frequently offered within the dominant discourse. This was a way to naturalize many human behaviors and institutions that were in flux or needed reinforcing. As the entomologist Vernon Kellogg (1908, 534) claimed of ants: "No insects are more familiar. They live in all lands and regions; they exist in enormous numbers; they are not driven away by the changes in primitive nature imposed by man's occupancy of the soil; they mine and tunnel his fields and invade his dwellings. And many things which man attempts they do more successfully than he does, and may be his teachers!" Systematic comparisons could become full-blown models that provided interlocking explanations for a wide variety of social institutions and practices.

Currently, the question of insect language and reason continues to be posed by entomologists and sociologists. The entomologist Zhanna Reznikova (2003) insists that there is evidence of complex interrelations in social insects, including the ability for learning and language. She claims that because of these capabilities, the interactions of social insects are political at a level similar to primates (2003, 5). Experiments conducted by Reznikova and Ryabko on large colonies of ants "have revealed a developed symbolic language in three ant species, which is probably even more intricate than in honeybee" (2003, 10). Reznikova makes a distinction concerning the individualized and politicized interactions she found in large colonies, claiming that, "The most important difference is that in species with small colonies, social life is based on anonymous contacts, i.e. these ants do not identify each other individually. The distinction between individualised and anonymous inter-relations in ants is comparable with Eible-Eibesfeldt's (1989) distinction between individualised and anonymous human societies" (2003, 11).

In the nineteenth and early twentieth centuries, language was seen by many to be a distinguishing characteristic of humans. Later in the twentieth century, the debate was renewed by Karl von Frisch's discovery of a form of symbolic communication among bees known as the "waggle dance," which indicated the direction and distance of food resources or new locations. The sociologist Eileen Crist (2004) maintains that the discovery of this insect "language" was significantly complicated due to previous assumptions that social insects were incapable of cognitive abilities. As von Frisch's discovery of the "waggle dance" became widespread knowledge in the United States during the 1960s and 1970s, the idea of language was still difficult to accept because of these previous conceptions. "It was a battle about received assumptions regarding insect capacities, and a willingness or unwillingness to abandon those assumptions in the face of a phenomenon that profoundly undermined them" (Crist 2004, 9).

For Crist, the controversy over "bee language" is significant for human societies in that it alters our perceptions of the world and our place in it. Her reflections on the meaning of insect language mirror the earlier claims of Maeterlinck. Both Crist and Maeterlinck understand that by naming language and thereby intellect in social insects, "our position in this world" as Maeterlinck phrased it, becomes less unique (Crist 2004, 35; Maeterlinck 1901, 178). Accepting the consciousness and learning capabilities of insects does reflect on our larger ideas of sociality and the social models of organization, as entomologists and sociologists of the nineteenth and early twentieth centuries knew all too well.

Despite the differences or because of them, analogies proved useful for both entomology and sociology during the nineteenth and early twentieth centuries. They helped to establish questions about individuals, collectivities, and social structures. Analogies also were used in the work of establishing the boundaries of both disciplines because they could map out the distinctions along with the similarities, therefore demarcating the fields of study (Gieryn 1983). Initially, both fields with an investment in universal evolutionary or later overarching ecological laws used the analogies to encompass their analysis of both human and animal societies. Not until later in the twentieth century would sociology make a significant break and mark human organization as the exclusive area of study for the discipline. Until that time, both disciplines helped create hierarchical analogies exchanging the natural and social imagery of insects and humans.

6

Naturalizing Hierarchical Sociality through Discourse

Identifying the power relations within the earlier dominant discourse enhances the discussion of anthropomorphism and human/animal analogies being used in more recent discourse. This is especially true because some terms used to describe social insects remain the same in the current literature. These analogies between social insects and humans are not abstract and outside of social institutions but rather have contributed to reinforcing hierarchical social relations. The terms used to describe social insects derived from hierarchical social institutions and roles, and that hierarchical quality was retained as the social systems were naturalized. This legitimating loop between social and natural systems co-created class, race, and gender hierarchies, as the following analysis of passages from the dominant discourse reveals. Their intersection presents a thoroughly hierarchical social organization, which will be discussed after passages on class, race, and gender are framed separately so that each of these specific analogies can be better understood. The naturalization of this social creation came about as social theory used analogies to social insects and entomologists built on social terms to describe their behavior. The dominant discourse was comprised of both fields (and the popular literature that borrowed from this information), creating a legitimating loop of terms and concepts that became accepted as universal models for social organization.

HIERARCHICAL ANALOGIES OF CLASS/CASTE

A key characteristic of eusociality is a specialized division of labor, specifically a reproductive division of labor. The description of this division of labor has included classes or castes with titles that mirror human social roles. In social insect colonies, there is a queen (in termite colonies also a king) and a worker and a soldier class; for various species there also are other specialized job titles and descriptions. The division of labor has also been considered a hallmark of a "higher" form of organization among humans as well. Sociological explanation for modern society revolved around the creation of a specialized socioeconomic division of labor. As Charles Ellwood wrote in the *American Journal of Sociology* in 1901:

> Let us consider the case of the social insects—the ants, bees, and wasps—to bring out our point still clearer. As is well known, these animals exhibit a marvelous degree of organization in the groups which they form, the division of labor and the corresponding division of individuals into classes among them often surpassing that found in human societies of considerable development. From an objective point of view these groups of insects seem as truly societies as any human groups. (1901, 729)

The division of labor was described as a universal law that not only spanned social insect and human societies but could be found in whatever socioeconomic system prevailed. Therefore, the eusocial insect colony can be compared to any modern society with a specialized division of labor. Although variations existed in the manner in which comparisons and conclusions were made, the shared emphasis on the division of labor as demarcating sociality was always present. This universalizing assumption can be attributed to ideas concerning social evolution. Wheeler describes the division of labor as an evolutionary advance responding to functional imperatives. He explains that, "problems could not be solved without a physiological division of labor among the individuals composing the society, and this, of course, implied the development of classes, or castes. Termite society was therefore divided into three distinct castes, according to the three fundamental organismal needs and functions, the workers being primarily nutritive, the soldiers defensive and the royal couple reproductive" (1928b, 211).

    The following entomological text describes a particular colony of bees that are only semi-social, with the division of labor being viewed as less complex than in the eusocial species. It is clear that not just any work is a mark of eusociality; rather, the work must feature specialized classes and tasks. These semi-social bees are compared to "rustics" who are "born to labor," and their division of labor is compared to that existing among Irish working-class humans because they both are seen to be limited to using simple labor-saving devices rather than having a complex specialized division of labor.

> They divide their labour in a curious manner. A bee settles on a tuft of moss, its head being turned from the nest, and its tail towards it: with its teeth and its first two legs it divides and disentangles the filaments, and transfers them to the two middle legs; the second pair seize and push them to the third pair, and these thrust them as far behind the tail of the bee as they can reach, by which means the moss is advanced towards the place where it is proposed to build the nest, by a space which somewhat exceeds the whole length of the body of one bee; another bee, placed in a line with the first, receives the ball of material with its fore-legs, and like the first, transfers it the whole length of its body; and thus four or five of these insects, stationed in a row, spare time and labour in conveying the material for building, on the same principle that Irish labourers may be seen transferring their wheelbarrows from one to another. (Anonymous 1837, 84)

The idea of a class structure is seen to be a natural benefit for insect community life as individuals function in tandem with each other for the same overall purpose. There are never any signs of class conflict in this dominant discourse. Eimer's optimistic view exemplifies this position: "These different classes all work together for the good of the whole community in the wonderful manner known to all of us" (1890, 423).

The division of labor might denote a fixed "class structure" in the analogies, but the issue of plasticity complicated the matter. At the same time that some authors noted the extreme specialization found in polymorphism to be a mark of eusociality, others questioned the evolutionary advantage. Some degree of plasticity, described as a shared human trait, informed the claim of the sociologist Alfred Espinas that ants were a more advanced social insect than certain bees or wasps.

While wasps and bees of various species accomplish only a small number of tasks almost always similar, ants' activities are almost indefinitely variable to circumstances. Some dig, others sculpt, others build, others collect, many hunt, some harvest and stock, these take the sap of flowers, those feed on their corollas, here you see them become slave owners, there breed aphids, and all these diverse actions are susceptible of limitless modifications, depending on inherited tendencies and particular circumstances. As a result of this aptitude to modify the effects of their activity, the division of labor must be, in a certain anthill, much more advanced than in a beehive. Indeed, when examining a working anthill, one could determine several categories of working ants; but (and that is typical of an advanced organism) the division of labor is not rigid and only leads to organic modifications in a limited number of species. (1878, 367–77)

Once again the degree of adaptation was an important indicator of the level of conscious reasoning. W. C. Allee maintained that insects have rigid castes, but that some amount of plasticity would be found in humans. However, in making his point he describes the human working class as fixed, whereas individuals in the upper classes may vary and might even share qualities of the lower castes.

One of the characteristics of social life among the insects is the presence of castes . . . which perform different functions within the colony. With many social insects the division of labor has developed to such an extent that the animals which do different work have bodies that are more or less structurally appropriate to their principal tasks. The reproductive female has a greatly enlarged abdomen; the soldier grows up to possess large jaws and heavy armor or other protective and attacking devices; a worker may be larger or small or medium in size, according as its size will best suit for some of the varied tasks necessary for the life of the whole colony. The situation is greatly different from that among human social castes, where a member of the aristocracy may be as husky of body and as empty of mind as the most menial of the working caste. (Allee 1938, 250–51)

Allee is indicating that there are castes for humans, and he also suggests what might be typical characteristics of those in the menial "working caste." In stating that it is possible that an aristocrat might share those same qualities, he is also letting the reader know that usually the upper caste does not. The issue, however, is one of polymorphism rather than caste in Allee's view; it appears that plasticity is found more often in the aristocracy than in the "working caste." Allee further indicated that a fixed sexual dimorphism existed within the human division of labor (1938, 251).

Within the dominant discourse, the line between class and caste is sometimes not clear because just as entomologists were trying to ascertain this about social insects, so too were sociologists debating the idea of caste in humans (see chapter 4). Although the same term was applied to social insects to describe their differentiated roles, one distinction that was made for humans was that castes were only of a professional nature and not usually expressed in a physical polymorphism. Nonetheless, some references were made as to how professional castes could lead to a type of physical polymorphism. Conversely, in some social insect species, castes are not expressed in an extreme polymorphism. In addition, although plasticity was stressed as a trait particular to humans, descriptions of what appear to be race-, class-, and gender-based castes are found throughout the dominant discourse; evidently not all humans were viewed as being endowed with the same plasticity.

The entomologist William Morton Wheeler compared the development of caste in social insects to the professional castes of human beings (1928a, 308). These roles for insects were described as "rigid" by Wheeler and therefore different from human professional roles. Although he noted this difference, Wheeler also suggested a development of polymorphism among humans in their professional roles: "[T]he normal human individual, though born into his society with peculiar hereditary endowments, is nevertheless so generalized and plastic that he can be converted through education into a more or less efficient member of any caste or even of several castes (professions).... [A]fter exercising his profession for many years, he may acquire the behaviouristic idiosyncrasies or even the physical stigmata of his particular calling. Even the actor, whose profession consists in a neuromuscular mimicry of individuals belonging to a number of castes, eventually acquires a characteristic facial and postural expression" (1928a, 309). Human plasticity appears to serve in this case as a means for redirecting caste through training for particular professions. Once learned, physical and behavioral changes mark the human as being within a particular "caste" or profession. The end result is remarkably similar to the system of caste among social insects. Interestingly enough, actors, whose profession requires constant plasticity in imitating other castes, still morph into their own distinct caste.

Similar to Wheeler's conclusion are Mukerjee's comparisons of the division of labor between humans and social insects. He begins with the usual list of distinctions on how social insects versus humans develop castes, with humans decidedly more plastic in behavior (1940, 36–37). However, as Mukerjee continues, it begins to look as if there might be some loss of plasticity and physical changes in the specialized division of labor for humans also. "No doubt an ever-increasing specialisation in human activities has brought about an atrophy or sub-atrophy of man's sense organs and tissues. An increasing proportion of the population in modern urban communities exhibits congenital or acquired defects of vision, hearing and smell; while degeneration in the epidermal structures viz. teeth, hair and pigmentation is on the increase. . . . There is also greater activity of man's visceral nervous system and endocrine glands which result in the higher emotivity and greater criminality in the large bee-hive cities of modern times" (1940, 37–38). Mukerjee pairs changes in physiology due to specialization with a reference to life in the "beehive cities." In this way, he signals a similarity between the polymorphic specialization that occurs under both human and social insect division of labor.

The sociologist Emory Bogardus also used the term "caste" to explain the rigidity of roles that defined the self for humans in any given profession ([1918] 1920). Although he located the pressure of human caste upon a social rather than a physiological source, he made sure to indicate that the social pressure to remain in a certain caste was "justifiable" and very rigid, and that it should define the self. Emory Bogardus suggested that human castes are extremely rigid, seemingly more so than Espinas in his descriptions of ants.

Workers of social insect colonies are presented as having all sorts of tasks; what varies in the dominant discourse is how these tasks are assigned. Although the dominant discourse tends to present a rigid and highly specialized division of labor, for certain species or in arguments about insect intelligence, the ability to switch tasks is occasionally introduced into descriptions of the division of labor. Although some authors presented age-related division of labor for bees for the purpose of describing overall social organization, a typical description may read more like the account of the entomologist's Karl von Frisch:

> So while the queen bee lays the eggs, it is the worker-bees who look after them. Apart from that, the workers also see to it that the hive is kept clean and at the right temperature; they remove waste matter and dead bees, act as architects of the bee residence, defend the hive if necessary, and busy themselves with the provision of food as well as with its distribution. In short, they perform all the duties with which the queen and the drones do not concern themselves, so that a well-organized division of labour exists among the members of a colony. Indeed this is carried so far that various duties are divided up again among various groups of worker bees, one group having to look after the nursery, another after the food, and yet another after the defence of the hive. Thus, in a bee colony, the individuals are all interdependent, not one being able to exist by itself alone. ([1927] 1953, 3)

In the following entomological text, the author cites Huber's detailed observation of bees and, similar to the human societies described by Bogardus, there are white-collar workers and blue-collar workers complete with "architects" and "mere bricklayers" that appear to be part of the division of labor among social insects: "It has been already stated that the community of bees is divided into three classes—workers, males, and a female. Huber has found that there is a division of labour among the workers; one set of workers are finished architects, who plan and build the edifice—they at the same time are the nurses of the young; while the other are mere bricklayers and plasterers, who only bring the raw material, but do not give it shape. The former he calls the nurse-bees; the latter, wax-workers" (Anonymous 1837, 49–50). The workers are not identified as female, which adds to the focus on class distinctions in this description. Workers are subdivided in terms of class—one group provides manual labor, and the other group, labeled architects, evidently has the skills for planning and building structures.

In Karl Marx's arguments concerning conscious labor, he made a distinction between bee and human architects: "We pre-suppose labour in a form that stamps it as exclusively human. A spider conducts operations that resemble those of a weaver, and a bee puts to shame many an architect in the construction of her cells. But what distinguishes the worst architect from the best of bees is this, that the architect raises his structure in imagination before he erects it in reality" (Tucker 1978, 344). The difference Marx identified between human architects or weavers and bees or spiders was consciousness. This is important because Marx did not accept that a human division of labor should be modeled from the social insects as human labor was unique. He also recognized the sociopolitical nature of the division of labor and the attempts to naturalize a particular idea of labor.

The naturalization of the human division of labor was deeply embedded in the dominant discourse; having come originally from a social principle and then being placed back onto examples in nature to reinforce that social principle. Because the dominant discourse did not describe class conflict with the model of eusocial insect societies, this would be at odds with Marx's ideas about the division of labor as well. According to McCook (1909, 55) the model of social insect society could instead create an industrial utopia:

> From the beginning to the end there was no discord among them; no protests; no strikes, sympathetic or otherwise; no walking delegates or their insect analogues; no oppressing (or oppressed) contractors or owners. Indeed, there was no occasion for any of these frequent appendages of great modern structures whereon human workingmen—artisans, mechanics, and common laborers—are engaged. And yet the work was done, and on undertakings relatively many times greater, in the most perfect harmony, good temper and content of all. Is it possible for man to draw some lessons from this example of natural civics? Is it beyond hope that some goodly measure of such results may lie within the sphere of the practicable for our current organized society?

The efficient division of labor among insects was viewed in McCook's account as a system to emulate. His version points not only to a lack of conflict but also a lack of owners. The division of labor among social insects was used as an example of many ideas on the organization of labor in human society.

A naturalized division of labor can reinforce a particular socioeconomic structure from feudal to modern society (or postindustrial society). As seen in the earlier examples, many tasks can be either agricultural or industrial. Karl von Frisch offers a description of both at different points in his text:

> This sugar excretion, "honey-dew," is carefully collected by the ants and brought back to the nest; the providers of it are carefully cherished and protected from attack. Some ants even go so far as to carry their greenflies [aphids] down to the depths of their nest in the autumn, where they hibernate and are brought up again in spring: like a farmer who takes his dairy cattle out of their winter stables for the spring grazing. (von Frisch [1927] 1953, 163)

> In a colony of bees division of labour is arranged as thoroughly as in a boot and shoe factory where a number of hands are employed, each in a different capacity: one for cutting the leather, another for stitching the cut out parts on a machine, a third for hammering nails in, and so on. Each one by keeping within the strictly limited range of certain activity acquires a special skill. Something very similar takes place in the bee's workshop: here the various activities are distributed among various groups of bees to such an extent that even the foragers are subdivided into a group of nectar-collectors and one of pollen-collectors, each group devoting itself exclusively to its own particular task. (von Frisch [1927] 1953, 14)

Various governmental structures can be utilized in the analogies, along with appropriate tasks within the specialized division of labor. For instance, Eugène Marais describes one observation specific to a monarchy: "We also saw the changing of the guards. The new guards entered the palace cavity by a large opening nearly opposite the head end of the cell, and formed a second circle within the circle of the guards about to be relieved" (1937, 150). Eugène Marais was born in South Africa in a Dutch colony and was educated in England. Despite what Ardrey (1969) describes as Marais' very ambiguous relationship to England, throughout his work it is clear that Marais still held monarchy to be the norm for a governmental system. Others who were not influenced by social location under a monarchy described the division of labor in a "constitutional state," a republic, anarchy, or some type of socialism or communism.

Many authors depict a breakdown in the division of labor as something that would have a negative effect on the governmental or sociopolitical structure and encourage crime. Citing Réaumur, one entomological text notes some discord in an otherwise "civil society" including a description of bees dueling and robbing hives or bees on the "highway." Organized crime seems to be the final outcome: "In hives which are ill managed, and not properly supplied with food, the bees, instead of continuing a well-constituted civil society, become a formidably organized band of robbers" (Anonymous 1837, 75). Fabre calls the ant a "land-pirate" and seeks to display another side to the ant's reputation as highly organized worker: "The shameless beggar, who does not hesitate at theft, is the Ant" (1912, 8). Common names for certain species of ants and bees associate them with larcenous behavior; they are called respectively "robber ants," "robber bees," or "thief ants." Kellogg uses the common name of one of these species and elaborates on it to describe a symbiotic relationship as thievery: "Through their narrower passages, too narrow to be traversed by the hosts, the tiny thief-ants thread their way through the other nest in their burglarious excursions" (1908, 544). Forel (1904, 20) describes robber bees as having "acquired the habit of stealing the honey from the hives of strangers. At first the robbers display some hesitation, though later they become more and more impudent." And according to Mukerjee: "Among the insects we also find, as in human society, results of perversion of the caste system in the existence of paupers and robbers or predators within the species . . . . The instincts to feed, foster, and defend other individuals make the invasion of an animal society by parasites and guests easy. . . . There is a considerable number of guests in the nests of ants and termites . . . there are many parasites in human society" (1940, 33–34).

Along with these social problems, there are tasks of social control that result in analogs to human social roles. Descriptions of police, guards, sentinels, and soldiers abound in the discourse. McCook includes some of these roles among a list of others within the division of labor: "There are sentinels or policemen, masons or builders, foragers, nurses, and courtiers or queen's body-guard" (1909, 9). These guards, sentinels, or policemen maintain social control within the colony, and not even a queen is exempt from this control. As Wheeler described one particular situation, a queen "is 'arrested,' to use Santchi's expression, by a band of Monomorium workers, which tug at her legs and antennae and draw her into the galleries. Sometimes she may be seen to dart suddenly into the entrance of her own accord and is arrested within the nest" (1928a, 295). However, Evrard (1923, 100–101) described a more common scenario, with the sentinels or guards portrayed as utterly devoted to their queen and the colony: "They are suspicious by temperament; wide awake detectives, they are also evil-tempered janissaries: you will see them bristling upon the threshold like a spiked and impenetrable portcullis beyond the moat of a mediaeval stronghold. The watch is well-kept. It is impossible to take them by surprise, or to evade them. The stronghold is defended by a rampart of love, devotion and self-sacrifice. Death is nothing, when confronted in the endeavour to protect and preserve the inviolable sanctuary." Latter's "sentinels" who are "stationed at the entrance" (1913, 68) are the same "inflexible and irascible . . . police-women" described by Evrard (1923, 102). The sentinels or policewomen offer protection from outsiders and social control within the colony.

In addition to agents of social control, a specialized class of workers termed "soldiers" is involved in defense of the colony. Describing a group of ants called the "mound-makers" (*Formica exsectoïdes*), the U.S. entomologist Henry McCook compares their defense system to the "militia organization of our earlier frontier States," emphasizing that, "[i]ndeed, such is, in theory, the relation of all citizens of the American republic to the general government (1909, 11). Wheeler (1928a) described the professionalization of the social insect "military," claiming that, "the methods of defence may be very different in different species and accordingly we find in both ants and termites soldiers adapted to particular offices, or 'professions'" (187). In this professional army, some ants or termites even serve as "directors of the foraging commissariat" (187). Soldier ants are identified by a particular physical trait of larger mandibles, but they are still workers; the degree of specialization is debatable. Presenting soldiers as a distinct caste and elaborating on their role and tasks has been used to naturalize the organizational structure of human armies.

HIERARCHICAL ANALOGIES OF RACE

Connected to the portrayal of war and aggression in the discourse are descriptions of a group of ants termed "army ants." These descriptions do not refer to the ants termed "soldiers" that exist in most colonies, but rather to those found generally within the subfamilies of Ecitoninae and Dorylinae, which are portrayed as functioning as a complete army. These ants were termed "ants of visitation" by the indigenous population as well as some early naturalists and entomologists. Their behavior was considered useful in controlling vermin and insect pests as they came into a dwelling en masse and consumed these on a regular basis. The interpretation of this behavior changed over time, influenced by a colonial perspective. Sleigh (2003) places the descriptions of army ants firmly within a colonial context, emphasizing that the fear colonizers felt of the colonized became expressed in the term "army ants." A telling passage that supports Sleigh's contention may be found in McCook's account of a colonizer's early encounter with army ants, which were not yet so named:

> This readiness for hostilities and ferocity in attack have been noted and recorded often of the hosts of true ants that swarm along the pathways of travellers in the tropics. For example, Stanley speaks of the "belligerent warriors" among the innumerable species of various colors that filled the African forests; of the "hot-water ants," as his men not inaptly named them, from the smarting pain of their stings; and of the minute red ants that everywhere covered the forest leaves and attacked his pioneers so viciously that their backs were soon blistered. (1909, 191–92)

As sympathetic as McCook finds Stanley and his men in their painful experience with the ants, he also makes it clear that the ants were acting out of self-protection, not malicious intent. After claiming that the behavior was simply "natural belligerency," McCook conflates the ant behavior with that of the native tribes who also attacked Stanley and his men. The expedition is portrayed again as undeserving and sympathetic victims: "It certainly seemed as little reasonable as were the unprovoked attacks of the human hordes of cannibal savages that assailed his expedition in their crowded boats, as he made his way through the heart of the Dark Continent, along the mighty Livingstone River. The tribes of ants and the tribes of men were not unlike in the native combativeness that animated them" (1909, 192).

Not only did passages such as McCook's reflect the racism of colonizers and their reaction to indigenous populations of humans (and ants), but it also illustrated the suppression of indigenous knowledge. Rather than use the indigenous names for these ants—names that reflected indigenous interpretations of the ants' behavior—colonizers renamed the ants. Some authors in the dominant discourse cite Thomas Belt as the one who reported that the indigenous population of Nicaragua may have termed these species "army ants." However, nothing in Belt's original work makes this seem likely. Belt does state, "In Nicaragua they are generally called 'Army Ants'" (1911, 18). However, Belt was a native of England, and his closest connections in Nicaragua were with the businessmen living in the country who were from the United States and England. His descriptions of indigenous people displayed an extremely racist and colonial perspective. There is no evidence to suggest that Thomas Belt used indigenous terms, unlike Maria Sybilla Merian in her interactions with the indigenous population in Surinam.

With the exception of comparing the movements of nomadic tribes to those of army ants, Belt more often refers to their military organization: "The moving columns of Ecitons are composed almost entirely of workers of different sizes, but at intervals of two or three yards there are larger and lighter-coloured individuals that will often stop, and sometimes run a little backward, halting and touching some of the ants with their antennae. They look like officers giving orders and directing the march of the column" (1911, 21). In his detailed descriptions of the army ants' behavior, he also describes other positions, such as that of the scout who communicates information, a position that existed in the British army. Belt held a very high opinion of the "intelligent insects," whereas throughout his writings he reveals an extremely low opinion of the human natives. Belt's summary of "army ants" cannot be understood as having anything in common with his disparaging remarks about the indigenous population:

> When we see these intelligent insects dwelling together in orderly communities of many thousands of individuals, their social instincts developed to a high degree of perfection, making their marches with the regularity of disciplined troops, showing ingenuity in the crossing of difficult places, assisting each other in danger, defending their nests at the risk of their own lives, communicating information rapidly to a great distance, making a regular division of work, the whole community taking charge of the rearing of the young, and all imbued with the strongest sense of industry, each individual labouring not for itself alone but also for its fellows—we may imagine that Sir Thomas More's description of Utopia might have been applied with greater justice to such a community than to any human society. (1911, 25–26)

Belt believed that this insect social organization was superior to human society in general, considering it a veritable utopia. Their military skills were seen to exhibit a "high degree of perfection" worthy of emulation.

Although Sleigh asserts that the ants symbolize the colonized, what appears to be more prevalent in the dominant discourse was the parallel that was observed between the intent and actions of the ants and the colonizers themselves. Not viewed in an ecological perspective as "ants of visitation" that come regularly to clean dwellings of pests, the renamed "army ants" appear to systematically colonize. Forel (1928) cites Vosseler's 1906 account of army ants: "When they are established in provisional quarters and have occupied or built a nest, they set out from this central point to organize hunting columns in the neighbourhood, and these columns may extend for more than 100 or 200 metres" (184). Vosseler's account continues to explain that when the resources are fully utilized, the ants move on from the "invaded region" to colonize other regions (184). The ants' "military maneuvers" were usually described admiringly in the discourse, and descriptions of "strategies and tactics" only became more elaborate through the twentieth century. Forel cites Savage (1874) in an early description that marks this transformation of discourse: "[T]he main column marches forward as before in all the order of an intellectual military discipline" (1928, 190). Forel himself, by combining both indigenous knowledge and this Western information in his description of these ants, shows terms in transition. This can be seen in his chapter title "Wars of the Visiting Ants, or Dorylinae, Against Other Living Beings" (1928, 179).

The morphology of the insect itself began to be described more as a soldier ever battle-ready: "Even at rest and at home, the army ants are always prepared, for every quiescent individual in the swarm was standing as erect as possible, with jaws widespread and ready, whether the great curved mahogany scimitars of the soldiers, or the little black daggers of the smaller workers" (Beebe 1921, 61). The nesting behavior of these ants created a physical mass that Beebe presented as all-encompassing: "This chocolate-covered mass with its myriad ivory dots was the home, the nest, the hearth, the nursery, the bridal suite, the kitchen, the bed and board of the army ants. It was the focus of all the lines and files which ravaged the jungle for food, of the battalions which attacked every living creature in their path, of the unnumbered rank and file which made them known to every Indian, to every inhabitant of these vast jungles" (1921, 60). Belt claimed the difference in color noted by Beebe to be a particular division of labor "the common dark-coloured workers and light-coloured officers" (1911, 23). This mirrored the division of labor of white officers who commanded native troops that fought in the British army. Overall, the "army ants" in these descriptions took on qualities that became more parallel to the British army in its colonizing efforts, including enslaving the colonized. As Julian Huxley proclaimed, "The best-known military activities of ants are those concerned with the raids of the slave-makers on the nests of the related species from which they wish to steal pupae to be reared as slaves" (1930, 65). The author quoted below is actually referring to Latreille's *Polyergus rufescens,* sometimes known as the "Amazon ant." These small red ants are a species considered to be "slave-making ants." He also claims that they are "legionary ants," which are actually the various species found in the subfamilies Ecitoninae and Dorylinae and are also known as "army ants." The descriptions of movements for both of these ants are similar, which may be why they are often described interchangeably; however, "army ants" do not take pupae from another nest to be raised as their own. The connections that were made between war and slavery are brought out in the following passage:

> Connected with the subject of the warfare of ants is the history of a species of this insect . . . called, by Huber, the Amazon, or Legionary Ant, the *Formica rufescens* of Latreille. It is both warlike and powerful, and, unlike the rest of the tribe, its habits are far from being industrious. Enough has been said to show that the proceedings of some insects so nearly resemble human actions, as to excite our greatest wonder: but the habits of the legionary ant are still more surprising than the proceedings of the chiefs which we have just described; it is actually found to be a slave-dealer, attacking the nests of other species, stealing their young, rearing them, and thus, by shifting all the domestic duties of their republic on strangers, escaping from labour themselves. This curious fact, first discovered by Huber, has been confirmed by Latreille, and is admitted by all naturalists. (Anonymous 1837, 127–28)

Slavery

The terms "slave-maker," "slave," and "slave raid" are very prominent in the dominant discourse of the nineteenth and early twentieth centuries and still remain in use. This usage has provoked many analogies to human society, directly by comparison and indirectly through the use of the terms. Some authors then and now attempt to qualify the terms by stating that the ants' behavior is not comparable to behavior associated with human slavery, but they then continue to use the term without further distinction. Ants are the only social insects said to exhibit this behavior. Similar to the discourse on gender and the "marriage flight," entire chapters are sometimes found devoted to the topic of slavery in ants. Vernon Kellogg, a U.S. entomologist, remarked that, "[t]he most interesting . . . of the familiar American Ants are the "slave-makers" and their "slaves" (1908, 547).

Forel's chapter titled "Slave-Making Ants" provides a good starting point for contextualizing this terminology as he acknowledged the alternative term and the potential for the mismatch in analogy. Yet, he still claimed that it was acceptable to use the term "slavery" to describe ant behavior and proceeded to do so. In presenting the "slave-maker" ant species, Forel described it as "undoubtedly the most intelligent, that is, the most modifiable in its instincts, of all known ants" (1928, 119). He described this distinction by stating that although the ant might "steal the nymphs of weaker species in order to make slaves of them, she is at the same time perfectly capable of working on her own account. On occasion she can dispense altogether with these slaves, which are . . . a luxury for her; she delegates some of her domestic labours to creatures which she has not taken the trouble to rear, and this sets herself free to roam about. This aim can hardly be a 'conscious' one in the human sense of the word, but facts speak for themselves, as we shall see" (1928, 119). Unlike other accounts, Forel makes a point of denying any dependency on the slaves in ant colonies. He also does not attribute consciousness to the act of slave-making like that which exists in human slavery. It might seem that Forel would not use this as an analog; however, in his debate with another entomologist we gain insight into why he believes this difference is not enough to forestall the use of an analogous term or comparison.

> The term "slave" drew storms of protest from Huber. Ants have no slaves, he declared but only "auxiliaries." . . . [T]he ant in her nymph-sheath has no knowledge of the companions which have nursed and cared for her; she has, therefore, no knowledge of her formicary. . . . Accordingly, ants which are stolen in their nymph-stage are not conscious in any way of being slaves, and work by hereditary instinct. In this sense, Huber is right. But this makes them less like auxiliaries or voluntary servants; they are "unconscious slaves," something like the children born in America of former negro slaves, but they do not think about the fact as some of the negroes were afterwards capable of doing. Hence I will retain the word "slave," though these slaves are just as much the masters of their mixed formicary as the species which stole them in their infancy. (1928, 119–20)

The ant slaves are described as "something like" second-generation human slaves, yet Forel maintains that in the case of ants, no conscious thought is involved in their condition. This important distinction did not convince Forel to abandon use of the terms of slavery to describe ant behavior; instead, he attempted more detailed analogies.

Despite the obvious incongruities between human and ant slavery, Forel believed "slave" to be an accurate term. He also believed slavery to be evidence of the natural convergence between social insects and man, simply a part of what he described as "the conditions of life (slavery and cattle-keeping among ants and men)" (1928, 13). Quite often the example of slavery was used to argue for either evolutionary or ecological convergences between humans and social insects. Emerson (1943) proposed that there did exist such an ecological convergence between humans and social insects that included slavery. This convergence then lent itself to analogies between the two groups. "Some of the most remarkable social adaptations are found among the slave-making species of ants. Convergence is also amply demonstrated in social groups. Even among the insects, highly complex societies arose convergently several times, the most striking example being the similar organizations of ants and termites, while the complex societies of insects and men are predominately convergent and analogous" (1943, 15–16).

So, too, did Darwin universalize this behavior, using the instinct of slave-making as one of his arguments for the law of natural selection. He described one particular species observed by Pierre Huber as "absolutely dependent on its slaves; without their aid, the species would certainly become extinct in a single year. The males and fertile female do no work of any kind, and the workers or sterile females, though most energetic and courageous in capturing slaves, do no other work" (1859, 334). Darwin then chronicles other entomological observations of this species and concludes: "If we had not known of any other slave-making ant, it would have been hopeless to speculate how so wonderful an instinct could have been perfected" (1859, 334). J. Clark (1998, 254) claims that Darwin did not wish to use the example of ant slavery to naturalize human slavery. However, Darwin's own observations of "slave-making" included noting a difference in color between those he calls slave-makers and those he calls slaves. In describing a survivor of a "slave-raid," he humanizes the event: "one was perched motionless with its own pupa in its mouth on the top of a spray of heath, an image of despair over its ravaged home"

(1859, 337). Despite what appears to be a sympathetic stance toward the ant "slaves," Darwin naturalizes slave-making as a larger universal law toward evolutionary progress: "it is far more satisfactory to look at such instincts as the young cuckoo ejecting its foster-brothers,—ants making slaves,—the larvae of ichneumonidae feeding within the live bodies of caterpillars,—not as specially endowed or created instincts, but as small consequences of one general law leading to the advancement of all organic beings,—namely, multiply, vary, let the strongest live and the weakest die" (1859, 360). For Darwin, ants were inspirational evidence of natural selection, from observations of altruism as well as supposed slavery.

The legitimating loop of "evidence" found in the natural behavior of ants described through a social lens provided detailed naturalization of the hierarchical institution of slavery. Many accounts describe two separate species termed "slaves" and "slave-makers" coexisting as separate races, and the difference in color between slaves and slave-makers is made the most of in these analogies. "The slave is distinguished from its master by being of a dark ash-colour, so as to be entitled to the name of Negro,— an epithet now appropriated to the *Formica fusca*, or ash-coloured ants. Their masters are light in colour" (Anonymous 1837, 128). In the dominant discourse, ant behavior also became conflated with specific color. Darker, enslaved ants were discussed with terms such as "weak," "timid," and "eager to serve," whereas their lighter-colored masters were "intelligent" and "brave," among other glowing descriptors. McCook states the more obvious connection of color difference alluded to in these terms and analogies: "It seems an odd coincidence, in view of the preference for African slaves among men, that the ants most affected by the slave-makers are the dark species and varieties, particularly *Formica fusca* and its glossy-black American variety *Formica subsericea*. Both the Sanguine and the Shining slave-makers victimize these species more freely, perhaps than any others. This is due, doubtless, to their greater feebleness and comparative timidity, as well as their adaptability to associated service" (1909, 263).

The distinction of color is significant in its relation to the slavery analogies but certainly not restricted to this usage. Because of the more obvious reinforcement of hierarchical institutions and practices based on color, other recurrent themes in the discourse become more questionable as well. Evrard (1923, 350) makes the following distinction of bees in a hive, dividing them into racial categories: "As for the habits, the temperament, the character of the hive, these may undergo some modification. The temperament of one hive differs from that of another just as nation differs from nation among men and race from race," As an example of this, Evrard presents the following: "if the chosen drone were of the black race, or a Carniolan, or a Cypriote, the new blood would have infused into subsequent generations an uncouth and aggressive temper, an excessive love of swarming, and a fierce, savage nature, all unknown to the Italian Bee. Half-castes exaggerate both the virtues and the defects of their ancestors" (1923, 350–51). Similar statements about "races" of bees are very common, such as these by Anna Comstock: "In comparison with the Italians the black bees are inferior in many particulars," and "The Italians are far more 'civilised' than are the black bees" (Comstock 1905, 48). Ant coloring became a marker for race that in turn was compared and could be hierarchically ranked in the analogies between social insects and humans.

The use of the terms surrounding race and slavery were not just isolated but rather were embedded in the larger colonial worldview found in the dominant discourse. Passages that compare insect societies and stages of sociality with human groups implied a social evolution that progressed from "savage" to "civilized." Much of the explanation for the use of such analogies is encoded in these passages. Lubbock combines the slavery analogy and a very obvious comparison of social evolution: "In character the different species of ants differ very much from one another. *F. fusca* . . . the one which is pre-eminently the 'slave' ant, is, as might be expected, extremely timid. . . . The slave-making ant (*P. rufescens* . . .) is perhaps the bravest of all" (1882, 27). Elaborating on this idea of difference among species, Lubbock then compares the species to "stages of human progress" (1882, 91). *Formica fusca* (the "slave ant") happens to be a species that "retain[s] the habits once common to all ants. They resemble the lower races of men, who subsist mainly on hunting" (91). Whereas,

> species such as *Lasius flavus* represent a distinctly higher type of social life; they show more skill in architecture, may literally be

said to have domesticated certain species of aphides, and may be compared to the pastoral stage of human progress—to the races which live on the produce of their flocks and herds. Their communities are more numerous; they act much more in concert; their battles are not mere single combats, but they know how to act in combination. I am disposed to hazard the conjecture that they will gradually exterminate the mere hunting species, just as savages disappear before more advanced races. (1882, 91)

The "extermination" of one ant species by another is made analogous to the extermination forced on an indigenous human population under colonization. The power dynamics of colonization are hidden in what is presented as a natural and seemingly inevitable evolutionary progress. As previously discussed, social evolutionary schemes provided categories that fit in linear sequence despite the actual relationship to time or place. Lubbock rounds off his expedient categories of ants to comply with a stage theory of evolution for humans: "Lastly, the agricultural nations may be compared with the harvesting ants. Thus there seems to be three principal types, offering a curious analogy to the three great phases—the hunting, pastoral, and agricultural stages—in the history of human development" (1882, 92).

A social evolutionary perspective viewed "simpler" forms as a means to trace the roots of evolution in its complex forms. The assumptions of this method include the colonial categorization of simple and complex forms, and this assumption appears in the analogies to social insects, as the following passages suggest:

There are other sorts of bees, whose history is less imposing perhaps, but not less curious, than that of the species which has been just detailed: some of these live in societies; but their limited number, and the rudeness of their dwellings, when compared with the populousness, [sic] the order, and the architecture of the hive, make us feel that we are contemplating a village, after having seen a large metropolis. The contrast, however, is interesting; and a true philosopher, after he has marked the manners of a civilized kingdom, always finds his knowledge enlarged by the observation of the simpler habits of the rustic, or even the ruder customs of the savage. The *Humble Bee,* so well known to us, may be truly termed a villager. (Anonymous 1837, 82)

🐜   🐜   🐜

It is interesting to observe that effects analogous to those we have been considering in insects may also be detected in civilized man compared with the savage or barbarian. (Wheeler 1928a, 312)

🐜   🐜   🐜

But the societies of ants of different species behave differently. Some of them, like civilized peoples, have fixed abodes; their formicaries are, in a sense, cities in enlarging which they daily labour. Ants of several other species have only temporary abodes and may be compared with the Tartars. They bivouac, so to speak, and are always ready to quit a spot where they have tarried as soon as it ceases to afford the conveniences that led them to select it. (Réaumur 1926, 135)

The comparison between levels of social insects, other animals, and humans on an evolutionary scale was typical of the dominant discourse. Distinctions were made between "non-civilized" groups and "civilized" groups, with the latter being the pinnacle of inherited knowledge. As is sometimes the case in these comparisons, social insects fare equally if not better than some human groups on the social evolutionary scale. Social insects that dwell in what appear to be modern cities are civilized, whereas nomads living in villages, whether of the human or insect variety, fall lower on the evolutionary scale. Herbert Spencer, whose theory of insect evolution Alfred Russel Wallace had so admired, proposed that it was a difference in quality and fullness of life that distinguished placement on the hierarchical scale:

> So a worm, ordinarily sheltered from most enemies by the earth it burrows through, which also supplies a sufficiency of its poor food, may have greater longevity than many of its annulose relatives, the insects; but one of these during its existence as larva and imago, may experience a greater quantity of the changes which constitute life. Nor is it otherwise when we compare the more evolved with the less evolved among mankind. The differ-

ence between the average lengths of the lives of savage and civilized is no true measure of the difference between the totalities of their two lives, considered as aggregates of thought, feeling, and action. (1879, 14)

Spencer ranks the differences between savage and civilized in terms of levels of complexity, and this can be applied to insects and humans alike, fostering comparisons on a graduated scale. Spencer's social evolutionary methods were very influential in the dominant discourse of the nineteenth and early twentieth centuries. In the following passages by Franklin Giddings (1896) and Robert Park (1936), the human family unit, discussed in terms of primitive or natural state and a more evolved state, becomes compared to the social organization of insects:

> The coöperation of animals in fishing, hunting, and defence is a functional association, but it is not sufficiently differentiated or regular to be regarded as a social constitution. Perhaps the permanent and systemic social organization of some species of ants is an exception. Likewise there is not true social constitution in the lowest band of savage men, although there is much coöperation in such communities and the family is becoming in some degree an artificial brotherhood through the inclusion of adopted members. (Giddings 1896, 172)

> In plant and animal communities structure is biologically determined, and so far as any division of labor exists at all it has a physiological and instinctive basis. The social insects afford a conspicuous example of this fact, and one interest in studying their habits, as Wheeler points out, is that they show the extent to which social organization can be developed on a purely physiological and instinctive basis, as is the case among human beings in the natural as distinguished from the institutional family. (Park 1936, 13)

The analysis of social organization is believed to illustrate evolutionary laws and may favor social insects over certain groups of humans. In Giddings's assessment, "social constitution" may be present in ants but not in the group he calls "savage men." In a similar vein, Giddings claims that it is possible that some species of ants hand down "rules of toleration and alliance" intergenerationally, whereas only a basic element for this development is found among "savage men." In the stages of social evolution, he accords this type of knowledge to those in the stage of "barbarism," but he assigns the peak of knowledge transmitted from generation to generation only to those who are considered "civilized groups" (1896, 143).

Wheeler cites several sociologists such as Comte on the accumulative properties of human society and then places social insects higher than some human groups in terms of their economic organization. Although social insects display the ability to accumulate social organizational knowledge, this is still mainly driven by instinct, Wheeler believes; however, he also adds, "this must have been true also of the earliest stages of human society. On the material plane savage communities are still far from attaining to a stage as economically accumulative as that of the hive-bee or the higher ants and termites" (1928a, 133).

It would seem from the above passages that the criteria for the organizational structure of social insects had been placed on human organizations. This would then imply a biologizing of human society in an attempt to create a scale and definition of sociality similar to the one found in nature. However, social evolutionary scales flourished before these comparisons and before the scale of eusociality for social insects had been created. The terms used in the descriptions and comparisons were socially hierarchical terms. And while this exchange did serve to naturalize social evolutionary scales based on hierarchal ranking, the creation of a sociality scale for insects was also influenced by the human ideas of social evolution. Maeterlinck, in describing the evolution of sociality in bees, goes as far as to project onto humblebees the colonialist idea of the human anathema to "civilized man"—the cannibal:

> Where there is progress, it is the result only of a more and more complete sacrifice of the individual to the general interest. Each one is compelled, first of all to renounce his vices, which are acts of independence. For instance, at the last stage but one of apiarian civilization, we find the humblebees, which are like our cannibals. The adult workers are incessantly hovering around the

eggs, which they seek to devour, and the mother has to display the utmost stubbornness in their defence. Then having freed himself from his most dangerous vices, each individual has to acquire a certain number of more and more painful virtues. (1901, 33–34)

For human and social insect progress, the self-restraint of individualism and the attainment of moral virtues fit neatly into a higher status on the social evolutionary scale. The moral lesson is entirely a human construction in this evolutionary scheme.

The social construction of insect sociality was premised on colonial views of what defined civilization as the "highest" and all others as "lower" in the social scale. According to one author, social evolution to a higher "civilized" state is inevitable and the bees have achieved it before humans: "And then one day a great revolution broke forth. It did not at first affect the whole nation of Bees; it seized upon them gradually, just as civilisation seized and is still seizing upon the human race, of which it has not as yet completed the conquest" (Evrard 1923, 355). Can this colonial-inspired concept of social evolutionary hierarchical scales adequately explain social insect behavior and organization any more than it can for human society? The legacy of these scales remains in the very definition of eusociality and in the terms used to describe insect behavior and roles. This in turn continues to be a source that naturalizes scales of hierarchy in human social structure through the legitimating loop of comparing natural and social worlds.

HIERARCHICAL ANALOGIES OF GENDER

Within this colonial-inspired social scale, how does gender factor in the analogies? As already explored, the descriptions of class/caste and race lend themselves to elaborate analogies that reinforce hierarchical institutions. These analogies in the discourse also employ designations of gender roles. Within a social insect colony, there are males and females with assigned roles that are often compared to those involved in human institutions and practices. Mukerjee details this differentiation: "Both the ant-nest and beehive include three kinds of individuals: at least one queen mother, who has lost the domestic art, a number of short-lived males, and a crowd of workers who have lost their maternal functions"

(1926, 226). Within the discourse, some authors distinguished human caste from the physical polymorphism of social insects; the sexual dimorphism, however, was acknowledged as an exception. Allee (1938) states that the only true physical caste in humans is that of sex, and this is extensive, going beyond mere reproduction: "[T]here are differences in sexual behavior and responsibilities which are associated with the more fundamental distinctions of sex. Frequently, as in man, these differences have developed into fairly distinct behavior patterns for the two sexes, until each sex is practically a distinct caste, almost in the sense used in discussing castes among the social insects" (1938, 251–52).

Although there was a convergence according to most authors, especially in terms of sexual dimorphism, there were some challenges in dealing with the analogies of ant, bee, and wasp colonies because they were populated with primarily female members. The dominant discourse had various ways of presenting analogies about a predominantly female society. Sometimes the issue was sidestepped or channeled into a particular type of analogy, such as a monarchy. Just as with other terms, qualifications could be made that would allow the continued use of the male pronoun. Julian Huxley (1930, 59) offers an example of this: "The domestication of plant-lice [aphids] has reached such a pitch among certain ants that the masters—or, rather, mistresses—collect the 'cows' eggs in autumn." On the next page, Huxley then uses the terms "herd-master" and "herd-mistresses" interchangeably. While there were those who grappled with the female pronouns, some authors embraced the idea of a female-centered colony, inspiring an oppositional discourse, which will be addressed in the next chapter.

Another way that this issue was resolved in the literature was to describe a range of female roles that mirrored stereotyped human gender roles. The distinction between nonreproductive and reproductive females led some authors to not assign a sex to the female workers at all and simply describe them as "workers." But for those who did assign a sex to the female workers, much was made of the fact that the workers were not capable of reproduction; for example, the nonreproductive female workers were sometime referred to as "incomplete" or other terms suggestive of a human bias toward female fertility. Workers are seen as valuable because of the tasks they perform: "The future, as well as the creation itself of an anthill, therefore depends on the working ants;

although fecund females sometimes join in the work, it is on the sterile females that everything rests, since if they didn't retain the fecund ones, the anthill would perish" (Espinas 1878, 371). And Latter states, "In accordance with the great divergence of duties there are also very marked differences of structure between the fertile queen and her sterile industrious daughters" (1913, 104). However, although the working females may be seen as important to the colony, they are still lower in status because of their nonreproductive status (see J. Clark 1998, 258–59, on the connections between nineteenth-century concerns about the rise in unmarried women and the metaphors of nonreproductive social insects). As Pycraft's statements illustrate, the female workers are "doomed for the most part to perpetual spinsterhood," whereas, "[o]ne youngster in ten thousand may attain to a higher state, may if Fate wills, become a queen and a mother (1913, 280). Therefore the question arises as to why the workers accept their fate. The earlier analogies placed this in a larger context of service to the state that reinforced both class and gender hierarchies: "Among the humble-bees, for instance, the workers do not dream of renouncing love, whereas our domestic bee lives in a state of perpetual chastity. And indeed we soon shall show how much more she has to abandon, in exchange for the comfort and security of the hive, for its architectural, economic, and political perfection" (Maeterlinck 1901, 34).

As mentioned in the previous discussion of the historical interpretations of colony composition, a discovery was made that the colony contained a queen instead of a king. This was not a difficult transition for those wanting to reinforce a monarchy ruled by a queen as a natural type of government. Many terms and phrases that were directly correlated to a human monarchy were used in the analogies. The queen is described as "her majesty" (Marais [1832] 1973, 30), her dwelling as "the royal apartment" or chamber (Forel 1928, 357), and her movements as "walking slowly, almost majestically" (von Frisch [1927] 1953, 20). The relation of her "subjects" to her is portrayed as one of service: "As soon as he had tendered his morsel to his sovereign" (Marais 1937, 150). The "queen-mother ruler" was thought to be easily recognized as royalty by the workers: "[I]t is not improbable that of themselves they distinguish between a queen and a worker, since they pay homage to the queen. We may then conjecture, that in order to distinguish one queen from another, some communication must be made by the individual queen to her subjects that she is their sovereign, since where the antennae were amputated in two queens, both were equally well treated, while a third, who had the antennae entire, was instantly recognized as a stranger" (Anonymous 1837, 43). Sensing that the queen is indeed their "sovereign" and not a worker or intruding queen, the bees then pay respect to her leadership. As a group, they show devotion or follow her movements in and around the hive. "It is astonishing to notice how quickly they [all the bees] become aware that their 'queen' has gone into the hive . . . so soon as I had put her at the entrance to the hive she walked in, and immediately with one consent, like a regiment of soldiers turning to 'right-about,' the whole concourse turned their faces to the hive and trooped in after her" (Latter 1913, 98–99).

Besides the alleged devotion that the workers show toward the queen, their complete dependence on the queen is also noted in these accounts. Experiments that removed a queen from the hive illustrated the chaos induced through the loss of the queen, as described in the following passage:

> At first, when the queen has been abstracted, everything goes on well for about an hour; after this space of time, some few of the workers appear in a state of great agitation; they forsake the young, relinquish their labor, and begin to traverse the hive in a furious manner. In their progress, wherever they meet a companion, they mutually cross their antennae, and the one which seems to have first discovered the national loss, communicates the sad news to its neighbour, by giving it a gentle tap with these organs. This one in its turn becomes agitated, runs over the cells, crossing and striking others. Thus in a short time the whole hive is thrown into confusion, everything is neglected, and the humming may be heard at a distance. (Anonymous 1837, 37)

The above passage mentions the absence of the queen as a "national loss"; this is a deliberate use of a phrase intended to reference an analogy to humans. The workers' loyalty is portrayed as comparable to the reaction that dependent subjects of a queen in a human monarchy might have if she died. The intention of an analogy to the human monarchy is fully acknowledged, although with a disclaimer that it is not "exactly similar."

Naturalizing Hierarchical Sociality through Discourse

The differences are not elaborated on, and the motives that drive social insects in their behavior toward the female primary reproductive are given as analogous to motives that might influence humans to serve a queen.

> It would perhaps be incorrect to ascribe this conduct to motives similar to those which influence human agents. And yet it is difficult, if not impossible, to resist the impression, that although not exactly similar, they are at least analogous. These humble creatures cherish their queen, feed her, and provide for her wants. They live only in her life, and die when she is taken away. Her absence deprives them of no organ, paralyzes no limb, yet in every case they neglect all their duties for twenty-four hours. They receive no stranger queen before the expiration of that time; and if deprived of the cherished object altogether, they refuse food, and quickly perish. (Anonymous 1837, 36)

The comparison to nation-states is common in the contemporary discourse. Julian Huxley, after cautioning against comparing humans and social insects, entitles his second chapter "The Ant State" and then compares the colony to a human nation: "[T]he nest comes to have a multiplicity of egg-laying queens, and so becomes potentially as immortal as a human nation. If such a colony continues to grow it will reach an inconvenient size, its food-needs outrunning its communications. When this happens new nests may be formed, just as new settlements were formed by the Greek cities, a band of workers with one or more queens leaving the old home to found a new one" (1930, 23). Queens and nations are inextricably linked in the analogies between social insects and humans.

As ruler of a nation, the queen is endowed with a status of privilege and power. Latter describes the "privileges of the lawful 'queen'" as including the ability to "stay at home all day moving about the nest, as is the way with the 'queen'" (1913, 94). Many authors were not clear as to how the social insect queen communicated her commands to the workers. Eugène Marais describes the termite colony in much the same way that Wheeler and others would use the superorganism concept. He felt that the colony was one organism that was controlled entirely by the queen on an unconscious level:

> The individual worker or soldier possesses no individual instincts. He forms part of a separate organism of which the queen is the psychological centre. The queen has the power, call it instinct if you will, of influencing the soldiers and workers in a certain way, which enables them to perform collective duties. This power or instinct she transmits to all queens born from her. As soon as the queen is destroyed all the instincts of workers and soldiers cease immediately. She transmits this psychological power to the future queens just as she transmits to them the power of producing three infinitely differing forms of insect: the queen, the worker and the soldier. (1937, 125)

Marais believed that the most important phenomenon of the termitary was the role of the queen in promoting social cohesion. He believed that his theory and experiments were unique in showing that the queen operates as a brain of the termitary organism. In one exploration, Marais excavated a termite mound and likened it to a skull housing the queen's brain: "We succeeded in cutting first the palace cavity and then the actual cell of the queen in half without causing any undue disturbance in the community. We simply carved away a portion of the skull and there before us lay the living, functioning brain of the organism" (1937, 149). Similar to Marais, Espinas described the role of the queen or mother as endowing social insects with a "single conscience of the community" (1878, 350–51). The mechanism behind collective behavior was attributed to the queen, whose psychological stamp was seen as being imprinted on each of the workers. The power of the queen was tied to the ultimate control over all of the workers, evoking their unconscious obedience.

The Male or Drones

If the female queen has the highest status in a colony, the male has the lowest status in bee and ant colonies. They are usually termed "drones," mainly within bee colonies. The only function for the males is reproduction; they are not seen as useful the majority of the time, especially in contrast to the workers. "When the young males come upon the scene they are treasured for their one necessary function, then cast off. They are not specialized as soldiers; nature has forbidden that by denying them defensive weapons; for a like reason they cannot work. They are simply dependants—nature's *beau ideal* of the "gentleman loafer" (McCook 1909, 202). Wheeler finds it curious that the males do not seem to have evolved into the full sociality of the species; with few exceptions, the males are not an integral part of the colony. Similar to McCook's interpretation, however, Wheeler tends to attribute some social standing to the males based on their reproductive function. He claimed that with some species of bees and wasps, the male was "slightly socialized," as expressed in their attachment to the hive. "But even this spark of sociability is lacking in the males of most Formicidae although they may be treated with solicitude by the worker personnel. . . . [A]s a rule, in all social Aculeates the males . . . are merely tolerated for a time as so many parasites on the colony. Being necessary for fecundation, however, they cannot be regarded as good-for-nothings" (1928a, 134).

Wheeler's claim that because of their function in reproduction male social insects cannot be termed "good-for-nothings" is a bit optimistic as compared to most of the interpretations. Espinas claims that "[m]ales remain indifferent to the society's works" (1878, 371). More than indifferent, though, the males sometimes were considered almost parasitic, as "idle males, subsisting upon the fruits of the industry of others" (Anonymous 1843, 82). As Wheeler suggested, they are not tolerated for long, their eventual treatment shown in Maeterlinck's chapter entitled "The Massacre of the Male." The unflattering portrait of the males revolved around their nonproductive existence in comparison to the workers. "Indelicate and wasteful, sleek and corpulent, fully content with their idle existence . . . they feast and carouse, throng along the alleys, obstruct the passages, and hinder the work; jostling and jostled fatuously pompous, swelled with foolish, good-natured contempt; harbouring never a suspicion of the deep and calculating scorn wherewith the workers regard them, of the constantly growing hatred to which they give rise, or of the destiny that awaits them" (Maeterlinck 1901, 347–48).

The workers destroy the males sometime after the cycle of reproduction has taken place. This act of destroying the males is not often portrayed as female hatred toward males for their gender but rather as workers' hatred toward a society's idle dependents. Males' low status in the colony does not reinforce an idea that human males have low status (but see J. Clark 1998, 259, on the nuances of some comparisons that suggest human males have unearned status similar to that of the drones). The analogy becomes one of service and usefulness to a social or political entity and a rejection of those seen as dependent. Although the males are the same species living within same colony, they are termed an "idle race" and similar categorizations unrelated to male gender (Maeterlinck 1901, 354). The analogy became one of ridding society of its dependents, as illustrated in Pycraft's (1913, 286) suggestion that the social insect society should be a model for human social structure: "There are no State pensions for those that are past work, but a State execution instead. This is vastly more economical, and it may yet commend itself to some would-be social 'reformers,' who will doubtless contrive to make exceptions to the rule!" Critics and supporters of human social reforms alike used social insect societies to make their case. The colony might be viewed as the ultimate collective, but the image of a collective that practiced natural eugenics was presented with equal conviction.

Marriage and Motherhood

The swarm of social insects and their "nuptial" or "marriage flight" are closely connected. Karl von Frisch described the stages that led from the swarm to a "marriage flight": "Spring, the time of blossoming . . . is also the most prolific breeding time. In consequence of the fast growth of the larvae, the queen's diligent egg-laying leads to a quick increase in the numbers of bees, and thus to a rapid strengthening of the colony. But it does not immediately lead to an increase in the number of colonies as well, because each colony with its queen . . . represents a 'state' complete in itself, so that the increase in brood increases only the number of its citizens" ([1927] 1953, 27). The idea of starting a new colony is compared to a rather human idea of state and citizen. When the colony population reaches a certain density, the old queen and the potential new queens leave the colony with a group of workers to start new colonies. The swarming behavior of bees is the result of this migration to form a new colony (von Frisch [1927] 1953, 27). As Latter explains, "Under normal circumstances, the 'queen' which emigrates with the swarm is the reigning and therefore impregnated 'queen' of the hive from which the excited throng issues" (1913, 96–97).

The analogy to a "state" is maintained as the colony attempts to adopt a new "ruler." However, the fact that this ruler is female genders this process; a marriage flight becomes the next step as the new queens emerge. "Emerging from her cell as a virgin, she must first achieve her marriage flight before she can begin to lay eggs" (von Frisch [1927] 1953, 28–29). Various authors treat the actual mating with a range of detail, but they always include some reference to the idea that this is the most important moment of the female's life. There is an emphasis on the insect bride losing her virginity as a result of the marriage flight, and many authors include some imagery of innocence. Immediately following the marriage flight there is a significant change in the queen's life and mannerisms in contrast: "[S]he takes to the air to unite with a drone high up in the skies. After that she becomes a sedate matron, never to leave her home again unless a young queen threatens to dethrone her in the following year: in that case she will rush out through the entrance again, but this time in the midst of a newly formed swarm" (von Frisch [1927] 1953, 29).

Roles that were predetermined as queen, worker, and drone included analogies concerning the institution of marriage and motherhood. The analogies in the dominant discourse between social insects and human were written predominately by male sociologists and entomologists, which might be seen as a factor in the observation and description of female roles and behavior. Sommer (2000, 29), in her analysis of primatology discourse, found that science was used in the service of politics to reinforce patriarchal hierarchies. Interjecting social agendas into studies of nature provided biased evidence of innate gender hierarchies. In the scientific discourse on primate behavior and social structure, narrowly proscribed female roles were defined and naturalized. The gender bias appears to be very clear in the following descriptions of marriage and motherhood in social insect society. Expectations about what these roles entail for social insects are drawn from stereotypical descriptions of women in human society. In some cases, they also reflect the stereotypical human male roles of pursuit and married life.

Réaumur actually described the mating ritual of ants in fairly clinical terms compared to most. However, he still incorporated a stereotyped gender description that reinforced this idea of female "transformation" through a sexual surrender to the male. He compared one set of ants that were already impregnated with a group that "had not yet submitted to the embraces of the male." Continuing with what he felt was scientific proof, Réaumur observed: "What seems to indicate that they were to be regarded as virgins is that when I took them the tones of their colouration were feebler than they were some days later after I had kept them. . . . It follows from this that they were only recently transformed at the time of their capture (Réaumur [1743–44] 1926, 169–70). Maeterlinck (1901) also mixed clinical terminology into his descriptions; however, his use of flowery anthropomorphic terms that fit gender stereotypes of courtship was also typical. Male "suitors" pursued the elusive queen, competing for the chance to mate with her. The victor is described as acquiring emotional satisfaction from this encounter with the queen, as well as his inevitable death.

> [T]his plumed horde sallies forth in search of the bride, who is indeed more royal, more difficult of conquest, than the most inaccessible princess of fairly legend; for twenty or thirty tribes will hasten from all the neighbouring cities, her court thus consisting of more than ten thousand suitors; and from these ten thousand one alone will be chosen for the unique kiss of an instant that shall wed him to death no less than to happiness; while the others will fly helplessly round the intertwined pair; and soon will perish without ever again beholding this prodigious and fatal apparition. (297–98)

Having described the males' role in the mating ritual, Maeterlinck moves on to the details of the flight and the significance of this "marriage" for the female and the future colony. As the female flies to extreme heights, the male bees follow. Her flight appears to be motivated not only by practical considerations and physiological changes but also by adherence to "mysterious forces":

> [O]beying the magnificent law of the race that chooses her lover, and enacts that the strongest alone shall attain her in the solitude of the ether, she rises still; and, for the first time in her life, the blue morning air rushes into her stigmata, singing its song, like the blood of heaven, in the myriad tubes of the tracheal sacs, nourished on space, that fill the centre of her body. She rises

still. A region must be found unhaunted by birds, that else might profane the mystery. She rises still; and already the ill-assorted troop below are dwindling and falling asunder. The feeble, infirm, the aged, unwelcome, ill-fed, who have flown from inactive or impoverished cities, these renounce the pursuit and disappear in the void. Only a small, indefatigable cluster remain, suspended in infinite opal. She summons her wings for one final effort; and now the chosen of incomprehensible forces has reached her, has seized her, and bounding aloft with united impetus, the ascending spiral of their intertwined flight whirls for one second in the hostile madness of love. (1901, 304–5)

Not only does this passage reinforce very human roles of courtship, but it also includes an evolutionary message concerning reproduction and courtship. The "suitors" are being weeded out in the language of social Darwinism; the mating reveals the theme of the survival of the fittest. The "chosen of incomprehensible forces" is the mate chosen by the mystery of natural selection. The refrains of this theme can still be heard in descriptions of mate selection as a genetic predisposition.

Another occurrence that is often described in detail is the struggle between "competing queens." These descriptions also reinforce the "survival of the fittest" value for the role of primary reproductive. At times, new queens are seen to fight with rival queens for the "throne." As Latter (1913, 102) explained: "The hive from which a swarm has gone forth contains the residue of the stock of workers, and, as a rule, several 'sealed' queen-cells containing queen-larvae or queen-pupae. . . . It is not until the eighth day after swarming that the most advanced of these young queens hatches out. Usually, but not invariably, she, with the assistance of the workers, slays her younger rivals who have not yet emerged from their cocoons." Unlike Latter's account and others, Espinas views the rivalry between queen bees as inspired by maternal instinct for those already in the hive, and not in terms of attempting to fight for the role of queen (1878, 367–68). Either way it is described, this phenomenon usually involves the sense of suspicion, jealousy, and hatred of females for each other, over a temporary male partner and the role of primary reproductive. The queen considered "successful" embarked on a "nuptial flight" where immediately afterwards her mate dies and she returned to the colony to lay eggs. The image of division between females in this case is accomplished even without a permanent male reproductive.

Because males in colonies of bees and ants do not continue to "rule" with the queen, there are not usually any marriage analogies beyond the few minutes of the mating flight. In a termite colony, however, with a male and female primary reproductive that remain active in the colony after mating, analogies emerge surrounding a "royal pair," and the gender roles are expanded to include both queen and king. Marais (1937, 25) begins an account of what happens to the termites after mating: "They are now beginning the final search—they are house-hunting, and this the male leaves to his wife. It must be a good house, for they will live in it for a long time. And with the finding of their home and the digging of the front door we will leave the happy pair for a while." The analogy goes so far as to provoke the imagery of a human dwelling. The female is presented as more invested in choosing the dwelling; this is because it will become her permanent location. After the "honeymoon," the next stage will be pregnancy and motherhood, with the female termite being confined within colony walls.

For the female termite, a transition into motherhood is not necessarily a flattering one. Gone is Maeterlinck's "bride" who is as "inaccessible as a princess in a fairly legend"; instead, "the queen grows larger and fatter. . . . Her small neat body vanishes in increasing layers of fat until . . . it becomes an unsightly wormlike bag of adiposity. And to heighten the tragedy, her mate . . . appears to have discovered the secret of eternal youth. He remains as beautiful and active and young as he was on his wedding flight. But if you look at her, an immovable disgusting worm, it seems impossible to believe that she ever fluttered in the air on fairy wings. We could hardly blame his majesty if he began casting an eye at some other female a little less repellent" (Marais 1937, 27–28). The male's reaction to the change in appearance of his female mate is heavily anthropomorphized with the noble qualities of an ideal husband. This depiction included a moralizing analog of how a human marriage might aspire to the termite example of "love and fidelity." Marais (1937, 27–28) romanticized the devotion of the termite male, claiming that, "[h]is attachment to his queen seems to keep pace with her own growth. If you lay open the palace cavity, he rushes round in consternation, but always returns to her side. There is no question of saving his own life in flight.

He clings to her gigantic body and tries to defend it, and if the ruthless attacker so wills, he dies at her side. What a wonderful example of married love and fidelity, which can survive this terrible change of his beloved to a loathsome mass of fat!" It appears that not only the termite queen loses her shape; Julian Huxley describes ants that "grow into great egg-laying machines only excelled in bloatedness by queen termites" (1930, 64).

If the discourse can reinforce such human gender roles as courtship, fidelity in marriage, and the shame of fading looks, it can also illustrate female lessons in child rearing. Perhaps influenced by being unconnected to a monarchy, the U.S. entomologist Henry McCook transforms the queen into a mother, claiming that the workers pay "a tribute to motherhood, not to queenhood" (1909, 162). In "dethroning" the queen, he places motherhood upon a pedestal instead. Protecting the eggs of the primary reproductive female contains only the ordinary "divinity" attached to obedience to natural law. McCook (1909, 162) emphatically claims that, "anything like regard to sovereign state, or purpose to give or maintain royal honors, is wholly foreign from the situation. Reverence for motherhood is there, however—wholesome and protected motherhood, the essential fountain of communal virtue, vigor, and perpetuity. Are we losing from our own race the due reverence of that 'divinity doth hedge about' maternity? Woe to the nations or peoples, be they ants or men, in such estate!"

In the analogies of the queen-as-royalty, it might be expected that class would intersect with gender. In these accounts, the queen may lay the eggs, but she has nursemaids who tend to them. The analogs to "maternal duties" are then constructed using the actions of these nursemaids. Nurses are depicted as naturally wanting to serve the queen in raising the young and as having genuine feelings for those in their care. McCook describes this as a rather mystical duty that springs from a larger sense of what often was referred to as the "spirit of the hive": "Leaving the queen and her body-guard, let us follow the fortune of the egg. From the queen mother it is carried into a separate room, presided over by attendants who have received the not inapt name of 'nurses.' . . . They are on duty at that point for reasons satisfactory to themselves and to the secret but all-sovereign Spirit of the Commune, whose mysterious sway all freely obey" (1909, 163). Individuals who serve the queen are presented as being fulfilled on a conscious personal level as well as following an unconscious obedience to the higher authority of the group. Réaumur submits this caregiving to an empirical test to determine if the nurses are as selfless as they appear. To begin this investigation, Réaumur ([1743–44] 1926, 204) asks: "Do they love their species with sufficient tenderness and generality to be disposed to lavish the same care on the larvae or another formicary as they do on their own?" His experiment involved placing one hundred nurse ants into a beaker with a dozen larvae from an entirely different colony together with one larva from their own colony.

> They taught me very soon that it is the general good of their species that inspires them and that their passionate affection is not confined to the larvae of their own family. The larva belonging to the hundred ants was treated no better than the strangers. All were equally acceptable. They set to work to prepare them a lodging at the bottom of the beaker; and made a cavity to which all of them were carried. Several of these larvae showed by their size that they were of the kind that transform into large winged females. It remains, perhaps, to investigate whether they would lavish the same attentions on larvae that produce worker ants.

Despite his overall conclusion, Réaumur realized that the equal treatment of all these larvae may have been motivated by the status of the larvae. His initial findings, however, led him to believe that the ants were dedicated to the "general good of their species."

The maternal instincts of the social insect nurses are compared to human parental instincts, and the social insect nurses are seen to serve as role models: "These nurses evince for the offspring of another greater affection than many parents show towards their own children" (Anonymous 1837, 59). Julian Huxley, who maintained that there were differences between the occupations of humans and social insects, freely cited and concurred with another author's description of a gender role occupation. After the emergence of a young ant, he notes that: "It is not fully pigmented, and is still enclosed in what looks like a shirt but is really its final moult. This is pulled off by the nurses. 'When we see how gently this is done' says Büchner, 'and how the young creature is then washed, brushed and fed, we are involuntarily reminded of the nursing of human babies'" (1930, 21). Karl von Frisch extended this image in detailing the first job of the bee as "foster-mother" or nursemaid. He explained that they provided a food developed in their salivary glands for the larvae and termed this food "mother's milk" ([1927], 1953, 37). Réaumur reinforced the belief that the social insects are appropriate models for maternal care: "When the ants retire into their habitation . . . it is not for the purpose of repose. It is there that the mothers lay their eggs, there that the hatching young become the great object of the solicitude of the workers, which are their nurses; and among insects and perhaps even among men it would be difficult to find any that are more attached to their nurslings and feel impelled to take such pains with them" (Réaumur [1743–44] 1926, 173).

If there are analogies made to the superior "maternal instincts" and child care of social insects, there also are, albeit less frequently, references to what might be their questionable practices. Julian Huxley uses an analogy between social insects and neglectful parents by speaking of "guests" (i.e., types of beetles) in an ant colony who offer secretions to workers in order to be fed by them: "Sometimes the secretions of the guests must be possessed of new and overpoweringly attractive properties to the taste, so that the nurses even neglect their proper charges in favour of those aliens. We know, alas! of human parents who will neglect their children for drink, but this is as if a mother were to abandon her baby for the charms of gin exuded from the body of a changeling" (1930, 80). Julian Huxley also provides a commentary on children that is rather ambiguous as to whether it is a positive or negative comparison. "Some of the leaf-builders provide us with a very startling piece of behaviour. They employ child labour" (16). The comparison involves ants utilizing pupae ("children") as a device for weaving. In using the analogy of child labor, Huxley makes a distinction between humans and ants, but it is an informative one. "The differences between humans and ant behaviour are beautifully illustrated by these ants. In so far as the ants adapt their weaving movements to the form of the rent, they are exercising a low but definite form of intelligence. But the basis of the whole action, the utilization of the larvae, is instinctive" (1930, 16). Actually using child labor is presented as appropriate for both ants and humans in this case; the analogy naturalizes this behavior for human society. The fairly large distinction between an insect pupa and a child is not noted. The only difference presented is that humans make the decision to employ children through the use of reason whereas for ants this is mostly instinctive. Since the behavior is shared, the distinction has an evolutionary component; instinct establishes the naturalness of the behavior, whereas reason supports its conscious application.

In the dominant discourse of the nineteenth and early twentieth centuries, hierarchical analogies of class, race, and gender reinforced an entire system based on these interlocking positions. Through specific comparisons to human institutions, social terms and behaviors were attributed to social insects that projected racist, classist, and sexist ideology. In turn, once these biased descriptions became part of the scientific discourse on nature, they could be used to naturalize social categories. Along with the hierarchy-based analogies between social insects and humans in the dominant discourse, another discourse presented social and political alternatives by using analogies in slightly different ways. This alternative discourse provides historical precedents for some of the models currently proposed to describe the social organization of humans and social insects.

7
Alternative Visions of Insect and (Human) Sociality

The analogies between social insect and human societies in the dominant discourse reinforced human hierarchical social institutions by naturalizing their creation and maintenance. In turn, the social interpretations of insect societies reflected a projection of existing hierarchical human institutions. However, to use Helena Cronin's term concerning evolutionary ideas, an "alternative tradition" was created alongside the dominant discourse (1991). Different interpretations and assumptions led to analogies that served to legitimate the potential for egalitarian structures in human society. Some of this oppositional discourse also questioned the use of particular terminology or hierarchical sociality scales in entomology. In the light of newly emerging models and analogies, the history of this alternative tradition could provide valuable background information to those who would challenge the dominant understandings of sociality and organizational structure. I present representative samples from three main themes within the oppositional discourse: feminism, cooperation, and postcolonialism, as well as the more general oppositional discourse regarding terminology and hierarchical sociality scales.

THE MATRIARCHAL INSECT SOCIETIES

The first incongruity that might strike anyone encountering the analogies between social insects and humans in the dominant discourse would be the sexist depiction of female roles despite what appears to be the essentially matriarchal societies of most social insects. If stereotypical roles are a product of patriarchy, how is it that these roles apply to the "matriarchal" societies of ants, bees, and wasps? As a counter to these stereotyped roles, feminist interpretations did exist in the nineteenth- and early twentieth-century literature as a part of an oppositional discourse. While some authors were actively committed to feminism, others included only some of these nonstereotypical interpretations alongside more hierarchical images. At times, a female-based evolutionary model was used that, while not explicitly feminist, did end up being an oppositional appeal for a less male-dominated society. This alternative evolutionary model was often tied to a belief in a feminine, holistic quality of nature, a view held by some organicist theorists, such as Theodor Eimer and Hermann Reinheimer (Sapp 1994).

As an example of the more "female-based evolutionary model" type of oppositional thought, Radhakamal Mukerjee cited Malcom Lyall Darling on the evolutionary role of matriarchy: "I contend that the matriarchal system in animal life, being selfless, is a move toward the development of an ethical system" (1940, 47). Mukerjee then proceeded to agree with Wheeler that human males create a "violent ferment" in human society and thus can be labeled a social problem. This social problem appeared to be solved in societies of insects and other matriarchal societies by controlling the population by gender. "Many insect societies . . . maintain their organisation by reducing or eliminating the number of males which is subtly and carefully adjusted to the needs of food-supply and reproduction" (1940, 48). These societies are "peaceful and harmonious," although this also places them at a plateau evolutionarily (1940, 47–48).

The ability of the social insects to regulate the gender balance was an underlying topic in both the dominant and the oppositional discourse. Mukerjee was careful not to advocate for a fully matriarchal society as a model for humans; however, he pressed for the evolutionary advantages of matriarchal influence in "socializing" males. In the dominant discourse, the limited role of the male in the work of the colony translated into lessons of class and nonproductive workers. In the oppositional discourse, the specific parallel of gender was maintained and analyzed. Mukerjee discussed the socialization of males as necessary in achieving a stable society: "There is no doubt that not before long centuries of evolution partially tamed the human male that early human society could be integrated and stabilised. . . . Matriarchal or matrilineal social organisation, though not universal, has been a significant phase in the evolution of a sane and balanced communal life as displayed in the order and social solidarity of the hymenoptera" (1940, 48–49). This ecological emphasis on cooperation and the socialization of males is shared with W. C. Allee of the Chicago Ecology Group. Allee had used the exact same quote from Darling that Mukerjee had, and surmised that females were the more evolved socially: "It is the human female that is the highly social force with our species, and in this we are again similar to the others mentioned" (1938, 261). He also, similar to Mukerjee, used termites as an example of male socialization: "Among the social animals only the termites have fully socialized males. . . . Termites are lowly insects, but in this one trait they lead the world. No one knows how the socialization of male termites was brought about, and if we should learn their secret it probably could not be applied directly to human affairs" (1938, 261).

Aside from the alternative discourse's tribute to matriarchy, generally the queen bee was romanticized as a royal figurehead that governed her people; that is, until after her "marriage flight," when she subsequently settled down and became an "egg-laying machine" and a mother. In political analogies in the dominant discourse, the images of the queen usually did not match those of the king bee officiating within his Parliament (Merrick 1988, Ransome 1937). To create political analogies that were matriarchal in more than a symbolic manner would lie outside of the dominant understanding. The alternative depiction below comes from Henry McCook, an author who also employed some of the analogies typical of the dominant discourse. Some of his observations, however, were decidedly influenced by the feminism of the day. In an entire chapter entitled "Female Government in Ant Communities," McCook described the insect societies not simply as a female monarchy, but rather in feminist terms. The ants bring essentialized female characteristics to bear on otherwise male-dominated roles to form a specific type of sociopolitical system. McCook claimed that, "the government of these emmet [ant] societies, as with bees, hornets, and wasps, is really a gynarchy, or government by females. Our worker ants are veritable Amazons. Not only does the entire domestic control and service of community fall to them, but also those more virile acts (according to human standards) of war and public discipline and defence" (1909, 155). McCook gave credit to female traits for the more evolved "public administration" skills of social insects: "That there is a 'female temperament,' sharply distinguished from that of the male, is obvious enough to the student of emmet habits. That its dominance is advantageous to these organizations the natural history of the Hymenoptera attests" (1909, 156).

McCook framed the "government" of social insects within an essentializing feminist perspective. A "female temperament" was considered proven by the natural world of the ants; however, as found in the alternative tradition, these female traits were presented in a positive light. Believing that a female-oriented government was effective for social insects, McCook cautiously, in the manner of Mukerjee, suggested that this might be, at least in some measure, an improvement for human society as well:

> What would be the effect upon human societies should similar conditions prevail among them? As a speculative theory it is worth discussing, and one would hardly err in thinking that our public and official affairs would be greatly bettered could woman's temperamental view of things have wider influence therein, especially in their relations to the young. Our civil governments and their administration, from the township to the national capital, are almost wholly products of the male element of the race. The predominance of the female element, which one sees in ant communes, might not be desirable in our present stage of civilization, although it would be an interesting experiment in a county or even in a State. Such illustrations as the United States presents throw little light upon the problem, for the general conditions of society in the States that give woman the suffrage really differ little from those prevailing elsewhere. They certainly fall short far of the female status in an ant commune. One may safely think that a great deal more of it would be to our advantage. The fact to be especially noted is that among ants, as also among other insects, nature has built up upon the female organization, and not upon the male, the most remarkable and successful examples of social life and government known to natural science—the ant commune, the beehive, and the hornet's nest. (1909, 156)

Alternative Visions of Insect and (Human) Sociality

McCook certainly felt that there was an ideal model for "social life and government" to be found in social insects; however, he seemed to suggest that it was a bit too soon in terms of evolutionary progress to have a full-fledged matriarchal government in human society. For those who at the time may have felt that the United States, in granting women suffrage, was somehow close to this realization, McCook adamantly dispelled this notion. It is clear that in McCook's estimation, human female suffrage did not approach the equality found within the ant colony.

McCook's views on feminism were not isolated, but were rather a part of the larger social reform movements that emerged in opposition to the hierarchical institutions of the time. Both naturalists and sociologists embraced alternative interpretations of nature and contributed to analogies that traded on this alternative view. As previously discussed, ideas concerning evolution were not monolithic, and these dovetailed with other issues such as feminism. As Fichman (1997, 110) attests, many prominent naturalists and sociologists were a part of this larger alternative tradition, including Alfred Russel Wallace and Charlotte Perkins Gilman. Sexual selection was framed in a socialist and feminist manner; natural evolution as seen through a reform orientation indicated a different path for social evolution. Within this alternative tradition, Gilman used analogies involving social insects to present what she envisioned would be less exploitative sociopolitical institutions. In the following example, she used the essentialist feminist argument common in her day; after she described male government as originating from hunters and fighters who "take all [they] can get," she presented the opposite vision of a potential peaceful, cooperative female government. She designated specific female traits that might be responsible for bringing about this improved governmental vision: "Government by women, so far as it is influenced by their sex, would be influenced by motherhood; and that would mean care, nurture, provision, education. We have to go far down the scale for any instance of organized motherhood, but we do find it in the hymenoptera; in the overflowing industry, prosperity, peace and loving service of the ant-hill and the bee-hive. These are the most highly socialized types of life, next to ours, and they are feminine types" (1911, 189–90).

A female-centered government based on the model of idealized motherhood was the vision for human society Gilman proposed. Social insects served as a living example of this type of government, and to Gilman the benefits of such a matriarchal society were obvious. To make her case that the female model had something to offer, Gilman initially presented the image of a fully matriarchal government model. However, she then shifted back into her humanist stance and declared that human society was composed of male and female; both have something to contribute and should be allowed to serve together to form a different and progressive type of government (1911, 189–90). This use of the matriarchal social insect colony as a model is shared by other writers only up to a certain point in the alternative tradition, as shown in the work of Mukerjee and Allee.

Also similar to other authors, Gilman posits the evolutionary predominance of the female as a universal force. Using this alternative concept, she attempts to put female subjugation in a larger comparative frame: "What neglect of faded wives can compare with the scorned, unnoticed death of the drone bee, starved, stung, shut out, walled up in wax, kept only for his momentary sex-function, and not absolutely necessary for that!" (1898, 134–35). The image of the male in this case is presented as more pitiful in order to raise up the female within an evolutionary perspective. Gilman is attempting to downplay women's subservience as the result of natural law, presenting it instead as a temporary social setback that will be overcome. She proclaimed that, "The female has been dominant for the main duration of life on earth" (1898, 135). In this statement, Gilman included the human and non-human female. In keeping with the evolutionary perspective, she also used terms that describe the female as "more socialized" or "more evolved" than the male.

The feminism of the entomologist Auguste Forel included extensive citation and an entire chapter of his book devoted to the experiments of a female colleague, Miss Adèle Fielde. He used her work and the example of ants to promote gender equality: "I have shown you . . . the significance of the most important experimental work of Miss Adele Fielde. . . . [Y]ou must needs join me in admiring her perseverance and exactitude, her highly conscientious methods, and the clear and acute judgment she displayed. This alone should convince you, if you were ever in doubt, of the desirability of votes for women! Even though there are many human imbeciles in both sexes, and though genius is rare on the whole, the male has no monopoly therein. Think of the ants and throw away your pride" (1928, 52). The model of an ant colony and the ability of females to conduct quality research provided an argument for gender equality within human society. Forel fell within the larger group of social reformers who advocated for class egalitarianism and feminism, believing that progress could be achieved in these realms.

These reforms included not only votes for women but also social and economic independence, as Gilman aptly argued for from her standpoint as a Victorian white, middle-class woman of the time. Women of her class were relegated to a separate sphere and forced to be dependent based on the "scientific evidence" that included biological arguments of women as the weaker sex. As is typical of Gilman's theoretical argumentation, she countered with oppositional interpretations of biology:

> Because of her maternal duties, the human female is said to be unable to get her own living. As the maternal duties of other females do not unfit them for getting their own living and also the livings of their young, it would seem that the human maternal duties require the segregation of the entire energies of the mother to the service of the child during her entire adult life, or so large a proportion of them that not enough remains to devote to the individual interests of the mother.
>
> Such a condition, did it exist, would of course excuse and justify the pitiful dependence of the human female, and her support by the male. As the queen bee, modified to maternity, is supported, not by the male, to be sure, but by her co-workers, the "old maids," the barren working bees, who labor so patiently and lovingly in their branch of the maternal duties of the hive, so would the human female, modified entirely to maternity, become unfit for any other exertion, and a helpless dependent. (1898, 18–19)

Gilman observed that women were not encouraged to work outside the home and that the argument for this was that they should expend all their energy on motherhood. Using biological analogies to social insects as a less desirable model in this situation, Gilman highlights the complete dependence of the queen bee "modified to maternity." Instead, she asserts that human females are not like bees, in that they are not biologically determined and "modified" to their maternal role. Even while being confined to the home because of this role, the reality was that women did a lot of "extra-maternal" work but were denied the economic independence that might go with that if the work were performed outside the home.

Gilman not only used her own analogies to social insects to assert women's independence, but she also publicly dismissed the analogies to social insects in the dominant discourse that reinforced women's dependence.

> In *The Forum* for November, 1888, Lester F. Ward published a paper called "Our Better Halves," in which was clearly shown the biological supremacy of the female sex. This naturally aroused much discussion; and in an answering article, "Woman's Place in Nature" (*The Forum*, May, 1889), Mr. Grant Allen very thoroughly states the general view on this subject. He says of woman: "I believe it to be true that she is very much less the race than man; that she is, indeed, not even half the race at present, but rather a part of it told specially off for the continuance of the species, just as truly as drones or male spiders are parts of their species told off for the performance of male-functions, or as "rotund" honey ants are individual insects told off to act as living honey jars to the community. She is the sex sacrificed to reproductive necessities."
>
> Since biological facts point to the very gradual introduction and development of the male organism solely as a reproductive necessity, and since women are sacrificed not to reproductive necessities, but to a most unnecessary and injurious degree of sex-indulgence under economic necessity, such a statement as Mr. Grant Allen's has elements of humor. The opinion is held, however, not only by the special students of biology and sociology, but by the general public, and demands most careful attention. (1898, 171–72)

Gilman had a keen awareness of how much the dominant scientific discourse had been used to naturalize gender social roles. An important element of Gilman's theoretical strategy had been to question these biologically deterministic analogies, to present an opposing analogy, and then to argue for the human social construction of gender roles.

As alluded to in the above rebuttal by Gilman, the alternative discourse was also shared by others who, like herself, viewed nature as largely a reflection of egalitarian principles. Yet the work of social theory in this vein also insisted on a distinct role of social reform to bring about an egalitarian society. Lester Ward was another sociologist involved in this tradition and important in the formation of Gilman's work, so much so that she dedicated her book *The Man-Made World; or, Our Androcentric Culture* to him. This dedication illustrated the significance of oppositional discourse for those looking for an alternative interpretation of nature and society as Gilman profusely thanks Ward for providing "a new grasp of the nature and processes of Society" stemming from his gynaecocentric theory (Gilman 1911).

Ward's theory of life stressed the importance of female contributions to both natural and social evolution. Culling an example from social insects, Ward (1888, 274) claimed: "The superiority of the queen bee over the drone is only a well-known illustration of a condition which with the usual variations and exceptions, is common to a great natural order of insects." Not only did Ward embrace a feminist perspective that proposed more emphasis on matriarchal origins, he also pushed for reform on the level of cooperative ideals as put forth in his ideas of telesis. Ward viewed society as made up of both natural and social forces and felt that human progress could be socially engineered to be based on cooperative and egalitarian principles. While Ward tended to more often use analogies from the plant kingdom, others who shared in the theories and analogies of cooperation used socials insects as models.

SOCIAL INSECTS AS SOCIALISTS

One of the overlaps with the alternative discourse of feminism of the time was an argument against the ideology of laissez-faire capitalism justified by the competitive, "survival of the fittest" interpretations of nature. Alternative interpretations presented a more cooperative possibility based on a natural law of mutual aid. Those such as Kropotkin viewed the behavior of social insects in a different light and therefore emphasized cooperation, seeing this as a universal law that pertained to human society as well. Kropotkin (1902, 14) attributed the complexity of the communities of termites and ants to their cooperative nature: "Their wonderful nests, their buildings, superior in relative size to those of man; their paved roads and overground vaulted galleries; their spacious halls and granaries; their corn-fields, harvesting and 'malting' of grain; their rational methods of nursing their eggs and larvae, and of building special nests for rearing the aphides whom Linnaeus so picturesquely described as 'the cows of the ants'; and finally, their courage, pluck, and superior intelligence—all these are the natural outcome of the mutual aid which they practise at every stage of their busy and laborious lives." Cooperation did not require relinquishing individual freedoms, and in fact the key to evolutionary progress was seen by Kropotkin to encompass both conditions: "If we knew no other facts from animal life than what we know about the ants and the termites, we already might safely conclude that mutual aid (which leads to mutual confidence, the first condition for courage) and individual initiative (the first condition for intellectual progress) are two factors infinitely more important than mutual struggle in the evolution of the animal kingdom" (14–15). Alfred Espinas (1878, 376–77) also claimed that a rigid caste or class system in ant colonies did not exist. Therefore, because individual ants performed a wide variety of tasks, the division of labor in ant societies did not result in hierarchy or leadership. Espinas's ideas on the division of labor reflected his belief that cooperation was more common in nature and should be emulated by human societies (L. Clark 1981).

The idea of natural selection at the group level was a full-fledged opposing argument against the natural selection of the individual. Survival of the fittest was seen as the survival of those who exhibited cooperation. This assumption allowed for very different interpretations of the natural phenomena exhibited within a social insect colony. And this in turn led to analogies to human society that legitimated cooperation and not competition as the more common natural law. As Kropotkin elaborated: "[A]nti-social instincts continue to exist amidst the bees as well; but natural selection continually must eliminate them, because in the long run the practice of solidarity proves much more advantageous to the species than the development of individuals endowed with predatory inclinations. The cunningest and the shrewdest are eliminated in favour of those who understand the advantages of sociable life and mutual support" (1902, 17–18). Those who advocated for the more cooperative model did not deny that competition existed, but they felt that this had been overemphasized, which distorted the observations of natural behavior. This ongoing debate spanned from early opponents of Darwinian evolutionary theories to the beginnings of ecological thought as exemplified by William Morton Wheeler. Wheeler (1928a, 5) maintained that, "the behaviour of every animal, figuratively speaking, revolves about two axes, one of which is aggressive and individualistic, the other co-operative, or social. The Darwinians took the former behaviour largely for granted and greatly stressed it, so that the latter appeared to be exceptional and in need of a special explanation. At the present time one might more properly require an explanation of the solitary mode of life, so deeply are all who study animals in their complicated living environment impressed by their social or associative proclivities."

In *Biologica Symposia VIII*, a collection of essays based on the work of the University of Chicago's Ecology Group and published in 1942, Gerard claimed that, "[a] reemphasis on cooperation as a biological 'drive' is also apparent . . . as well as a rising tide of analysis of, even concern for, the position of science in modern civilization" (79). For those who espoused cooperation as the driving force of society, large-scale conflict as exhibited by war was seen as the antithesis to laws of nature and society. During World War I and World War II, there was a strong reaction against war, and many social and natural scientists searched for universal laws that would aid in the argument against the need for war. (See W. C. Allee 1943, "Where Angels Fear to Tread: A Contribution from General Sociology to Human Ethics" *Science* 97 [2528]: 519–25.) Using social insect colonies as natural examples of cooperation was one argument against the natural inevitability of conflict and war. However, scientists working in the alternative discourse had to first take on the interpretations found in the dominant discourse.

Eugène Marais, who had been greatly affected by the devastation of the Boer War, challenged the interpretation of war as described by Dr. Bugnion in Auguste Forel's book: "Dr. Bugnion discusses in particular the wars between the ants and the termites, with special reference to the origin of instinct. As would be the case in any tropical country, Dr. Bugnion saw many instances of attacks on termites by ants. He ascribes all the instincts and variations in form of the termites to this continuous state of warfare. I must state at once that I had practically no evidence of this ant warfare in Waterberg" ([1937] 1973, 110–11). Marais replaced Bugnion's account with a more cooperative interpretation. "[I]t is easy to get an impression of war, which nevertheless is based on inaccurate observation. There is no war; in fact most probably it is protection and friendship. This may be proved by anyone who cares to do so. If we break down a number of the smaller termitaries, sooner or later we come upon one which ants and termites occupy together" (110–11). Marais contended that the closer observation found that the analogy to murder or war was incorrect (111). Not only did Marais believe that the interpretation of war was inaccurate, but further, he believed almost the complete opposite, that the behavior was protective in nature.

Auguste Forel, who had included Dr. Bugnion's description of ants at war as an appendix to his book, nonetheless held a different view of the "ant war" analogies than did Bugnion. Forel, similar to Marais, Allee, Emerson, and others, viewed ant colonies in a manner that was influenced by a reaction and opposition to war. Although he studied and wrote about conflict between ants, he noted that this was different than the "insane wars between man and man, mostly waged in the name of national chauvinism, falsely termed patriotism," and observed that the conflicts between ants and termites were against other species as opposed to humans fighting their own species (1928, xvii). Although he did not embrace a version of complete passivity for social insect behavior, he nonetheless focused on the cooperative aspects within a colony. Forel was an avowed socialist and feminist. Comparing himself to the author of his appendix, he wrote of Professor Bugnion: "He is no more of a Bolshevist than I, but I believe in Socialism—and therein we differ" (1928, xix). Forel's vision for world peace took a page from socialism and the world of social insects: "If you want peace, prepare for it by disarming all men and organizing them in accordance with a true labour socialism of peoples that elect for themselves a supernational authority, supervised by them and responsible to them. We should accomplish this by means of an international alliance based upon that of the polycalic formicaries and combined with that of the mixed colonies" (1928, xix–xx).

Although Forel documented the "wars" of ants and termites, he also became aware of their cooperative behavior, sharing nests, "*somewhat like those of lodgers in the same house*" (1928, 105). Forel was so taken with this behavior that he set about to give it a new term, calling it "parabiosis." Marais had observed the same behavior and considered it a "friendship" and "protection" between ants and termites. Forel claimed this behavior as a new observation for entomology; it was certainly one that went against the dominant thinking of the time as characterized by discourse such as Bugnion's. Forel's moment of discovery was tantamount to a paradigm shift for his own research. "I stood thunderstruck at the thought of such a thing, which was then absolutely new in the annals of myrmecology. After due reflection I baptized it with the name of *parabiosis* (life side by side)." Forel then explains how this discovery would lead to a wider alternative perception: "You may imagine, dear reader, how this discovery opened my eyes to other cases, more or less analogous, to which I had not hitherto given enough thought" (1928, 105). Parabiosis then became part of Forel's analogy for human society under a universal formicary. After the heading "International Parabiosis of the Nations of the World," he wrote: "long before any of us the ants had realized the true universal fraternity in equality and liberty, in the form of *parabiosis* for communal work" (1928, 350).

Forel's parabiosis for humans could not compare to social insects exactly because human nature, according to Forel, required more measures of social control to keep it in check. "The hereditary social instinct of ants permits them to live without chieftains, guides, police or laws, in an admirably co-ordinated state of anarchy; human beings are absolutely incapable of doing this, and if they attempt as much as they at once fall back into such a triumphant state of brigandage that they are compelled to submit once more to the rule of chieftains. Such is the ancient tragedy of humanity, a thousand times repeated throughout history" (1928, 337). Although pessimistic that human society could achieve the level of self-control of social insect society, Forel believed that there would be a way to achieve cooperation. He felt that humans could, to a limited extent, emulate the higher form of social cooperation found in social insects. Forel (1928, 33) phrased the challenge in the following manner: "What must we do, then, to grow nearer to the ants and yet remain men?" Unlike some others in the alternative tradition, Forel's vision did not include anarchy or any type of self-organization for humans, and he believed this interpretation of nature as a social model to be mistaken: "The whole history of human races proves to satiety that we are absolutely incapable of living in the happy and well-co-ordinated anarchy represented by a formicary, and shows how misguided was the all-too-famous Kropotkine" [sic] (1928, 344). It would seem, according to Forel, that humans could only achieve some type of internationally organized socialism, whereas social insects could evolve into more leaderless forms of anarchy and communism.

Anna Comstock also referred to the organization of a bee colony in terms of socialism or communism. This appeared to be a form that was in its ideal state and was not compared to any existing human form. Comstock viewed the social insects as a model for how communism or socialism (terms she uses interchangeably) might be best understood (1905, 6). The perfect state of socialism consisted of a move from qualities emanating from the individual alone to those shared by the entire community. This was accomplished over time and appeared, from Comstock's description, to be a type of evolutionary advance:

> The bees and their relatives are the most intelligent and consistent socialists that have yet been developed in this world; and, through studying their ways, one may discern with startling clearness how the perfect socialism grinds off all the projecting corners of the individual until it fits perfectly in its communal niche. In the hive individual traits, as exemplified by kindness, selfishness, love and hate, are moved up a notch in the scale and characterise the whole community, even though they are eliminated from its members. If one is a social philosopher, he may become very wise, indeed, by studying the results of the laws of socialism which have been executed inexorably through countless centuries in the bee commune. (1905, 6–7)

Thomas H. Huxley (1894, 24–25) also believed that the social insects exhibited a communistic form of government, which he equated to the human counterpart. A "cosmic process" of organization applied to both social insects and humans. "Social organization is not peculiar to men. Other societies, such as those constituted by bees and ants, have also arisen out of the advantage of co-operation in the struggle for existence.... The society formed by the hive bee fulfils the ideal of the communistic aphorism 'to each according to his needs, from each according to his capacity.' . . . Queen, drones, and workers have each their allotted sufficiency of food; each performs the function assigned to it . . . and all contribute to the success of the whole cooperative society." Huxley's sense of communism ruled out individual freedoms and placed more emphasis on the collective good over the individual: "Each bee has its duty and none has any rights." The socialism or communism attributed to social insects evidently adhered closely to the version that a given entomologist or sociologist chose to present.

Henry McCook (1909) not only attributed communistic qualities to social insect societies, but also alluded to the self-organizing properties that many entomologists currently believe to be a determining force in the division of labor:

> Every ant is a law unto itself; and in every individual the self-directing faculty is well-nigh perfect. There is no private property. All citizens are equals—absolutely equals in ownership of the communal property and in the use of, the authority over, and the service and responsibility of the same. All serve, save natural dependents; but all apparently are free to choose the quality, the period, and the amount of service. There is no visible head, no representative class or body within which the control of the common-wealth is embodied; and yet, by some occult force hitherto unknown to men, all the beneficent effects of government are wrought out with the regularity and precision of an automatic machine. (303)

The mechanism remained a mystery, but many authors believed that there was some amount of self-organizing in social insect colonies. This self-organizing form made a convenient analogy for the alternative tradition, usually being correlated to anarchy, communism, or various stages of socialism.

SOCIAL INSECT COLONIES AND POSTCOLONIALISM

Oppositional discourse of a postcolonial nature is extremely rare in the dominant discourse of the nineteenth and early twentieth centuries for several reasons: colonial power was at its height; most of the writers were from colonizing countries; and ideas of social evolution permeated the discourse. Science contributed to the ideology and practice of imperialism. The preconceived notions of a colonial worldview were embedded in the observations and descriptions of nature, and were offered in turn as proof of the more advanced powers of the colonizing nations and peoples (Greene 1981, 121). The legitimating loop of social and natural comparisons by and large reinforced the hierarchical structures and practices of colonialism.

There are only a few examples of what might be considered a postcolonial sensibility found within the dominant discourse of the nineteenth and early twentieth centuries. As previously noted, Radhakamal Mukerjee affords some critiques on Western conceptions of India, while remaining influenced to some degree by the colonial worldview. Mukerjee did provide a unique perspective on the behavior of social insects in that swarming was likened to colonization. Although not launching a full criticism of slavery, he sometimes described slavery as abnormal: "Other instances of perversion of behaviour are afforded by slavery or dulosis among some species of ants whose workers raid the nests of other species carrying off to their own nests pupae from which workers subsequently develop" (1940, 34). Other authors were inclined to describe slavery as destructive only in the extreme case of a particular species where the slave-makers became too dependent on the slaves and lost all autonomy, which was seen as undesirable for the slave-makers' evolutionary progress. This is vastly different from Mukerjee's unqualified identification of slavery itself as a "perversion."

In his analogy to slave-making ants, Henry McCook (1909) first reviewed the history of human slavery and made distinctions between slavery that focused on class bondage and slavery that was racially based. He maintained that the slavery of Rome and Greece, as compared to the slavery of colonialism, did at least allow former slaves to become full citizens, and that, unlike American slavery, it was not based on racial castes.

> Our British forebears, to whom we owe our views of both civil liberty and chattel slavery, were at one with all Europe in holding Africans as the lawful prey of white men. . . . One needs this bird's-eye glance at this phase of human society as he takes up a somewhat analogous feature of certain ant communes; for our

conception of ant "slavery" is colored by the current meaning of the word as from our own use and wont. It is not, indeed, an inapt term as applied to emmet communities, if one regard the usage of men in the whole course of social history; but it is a different thing as interpreted by one's preconceptions of slavery as lately existing in the United States. (262)

McCook believed that when the term "slavery" was used to describe ant behavior, the vision of human slavery in the colonial period was the image that came to mind. Like some others, McCook described the analogy to colonial-period slavery as an inaccurate comparison that should not be applied.

As postcolonial theorists point out, the relationships within colonialism are quite complex. One example of this comes from the standpoint of the Afrikaner in relation to imperial science. (See Dubow 2000 for more on this dynamic.) Eugène Marais was in a unique position to be critical of the colonial worldview. Marais was a Afrikaner who happened to be in London when the Boer War began, which resulted in his being classified as an "enemy alien on parole" (Ardrey 1969). At the end of the war after an escape from Britain, Marais attempted to get supplies to his fellow Afrikaners but was too late. Ardrey (1969, 6) claims that Marais "never forgot" the horrors of the Boer War. This event and his social location may help to explain Marais' position in the alternative discourse.

After the war, Marais subsequently isolated himself in the wild to study animal behavior and published primarily in the Afrikaans language. Within this larger context, therefore, Marais' critique of Dr. Bugnion's observations of termites began with a telling preface: "In connection with this riddle, I want to show how modern European learning handles cases of this kind, and the explanation it finds. I am able to do this through a correspondent, personally unknown to me, who sent me a monograph written by Professor Dr. Bugnion of the University of Lausanne" ([1937] 1973, 110). After debunking Dr. Bugnion's findings of termites in Ceylon, it is clear that Marais viewed him as a visitor who did not know the country well enough, which led to "unsound conclusions" ([1937] 1973, 112). His prefatory remark was meant to be critical of the methods and preconceived notions of "modern European learning."

His corrective observation spoke of his own sense of intimacy with the country he felt to be his own. "Much time is necessary to study even a single phenomenon of termite behaviour in a dry country like South Africa" ([1937] 1973, 111–12).

Imperialist expansion resulted not only in the colonization of people, land, and natural resources, but also in the appropriation of indigenous knowledge. Ritvo (1997b, 336) points out that, "naturalists in the mother country automatically claimed the right to classify colonial plants and animals—their subject in more than one sense." In the case of entomology, colonized people were brought into the scientific imperialist project by being pressed into collecting insects to send back to Europe to be studied, classified and renamed by Europeans (Harries 2000). As a result of this colonizing process, the colonizers eventually claimed that only their schemes for classifying social insects represented legitimate scientific knowledge.

However, because this colonizing process of classifying and renaming was gradual, some of the earlier discourse retained indigenous terms. Although the use of these indigenous terms certainly cannot be described as part of a conscious oppositional discourse, it did temporarily preserve indigenous knowledge about social insects and indirectly highlighted how a particular relationship and different worldview provided an alternative explanation. Réaumur retains the terminology and description of what would later be termed "army ants." He even cites Maria Sybilla Merian, who would later be criticized for using indigenous terms and descriptions in her work:

> Father du Tertre and also Father Labat . . . report concerning the ants of Martinique what Mlle. Merian tells us about those of Surinam. For two or three consecutive days troops of these insects present themselves to visit the houses. They arrive in files so broad and so dense, so continuous and so long, that it is vain to endeavour to oppose their progress. But experience has shown that, far from coming to cause annoyance, they come only with good intentions. Father Labat assures us that all the doors are willingly opened to them on their approach. They overrun one after the other all the rooms of the house that they enter, from cel-

lar to garret, and forage in every nook and corner of every room. Any room they choose to enter has to be abandoned to them, so that the human occupants retire. In the meantime they kill all the insects they encounter and clean out the house, but the most useful of their occupations is the destruction of the cockroaches. They would render an important service to the dwellers in every house if they could deliver them entirely from these pests, because there are many complaints in regard to them. When these ants have thoroughly overrun the house from top to bottom they leave it to enter that of a neighbour. They have been called "ants of visitation" and they deserve the name. They do not become a nuisance, because their visits occur only once a year and are not, therefore, too frequent. (Réaumur [1743–44] 1926, 186–87)

"Ants of visitation" were not described in militaristic terms in this passage; instead, an attempt is made to describe their actions in relation to the indigenous population. Indigenous terms and observations were sometimes noted alongside a Western term in the transition to the exclusive use of Western terms. However, this was generally accompanied by the qualification that these were native terms and therefore not given the legitimacy that is suggested by the quotation from Réaumur's work.

Another aspect of the alternative discourse that could be viewed as opposition to colonial thought was a rejection of hierarchal social scales. Early on within the entomological discourse there existed a critique of privileging the study of certain insects over others. The bias of a hierarchical social scale was discussed as problematical in its exclusionary effect:

Bees, ants, and wasps are the familiar Hymenoptera. They are the "intelligent" and the "social" insects, and therefore seem, of all the insect hosts, those living the most specialized or "highest" kind of life. As intelligence and social life are precisely those characteristics of our own which most distinctly set us off from other animals, we are quick to appreciate the worth of similar attributes in the "ant and bee people." But in actual degree of specialization of instinct and behavior the performances of solitary wasps and bees are little less wonderful than those of the social kinds, and the amazing character of the life-history of many of the obscure and unfamiliar parasitic and gall-making Hymenoptera ought to incite as much interest and scientific curiosity as the marvels of the bee community. (Kellogg 1908, 459)

C. Lloyd Morgan placed equal importance on solitary and social forms as survival strategies. He disagreed with the rigid distinctions on the insect hierarchical scale and then compared this to the human social scale, noting that basic human qualities such as the human senses were shared by "civilized" and "savage" alike. Morgan takes issue with Kropotkin's overemphasis on sociality to illustrate that any social insects should not necessarily be seen as evolutionarily more advanced: "The assertion that the fittest are the most sociable animals, that sociability appears as the chief factor in evolution, and that unsociable species decay, is not likely to be accepted without qualifications by zoologists. What grounds have we for saying that the solitary wasps are less fit than the social wasps? Each has a fitness according to its kind" (1895, 229). Kropotkin argued for change in the definition of eusociality, viewing cooperative, not competitive, sociality as the most adaptive. Morgan took the challenge another direction. If, according to Morgan, evolutionary advantages might be had for both social and nonsocial animals, a ranked scale should not be necessary.

For W. C. Allee, the force of group pressure on interaction made it more difficult to understand the extent of social behavior. He addressed this complexity by blurring the lines between categories of any type of social scale: "Such considerations serve again to emphasize the difficulty of drawing a hard and fast line, or even a fairly distinct band between social and sub-social living" (1938, 258–59). Regarding both Mammalia and Insecta as having the highest degree of sociality, he nevertheless wished to claim sociality at all levels, similar to Kropotkin's emphasis on this as the key to social evolution. Allee argued that, "since no one has yet demonstrated the existence of truly asocial animals it is impossible to define the lower limits of sub-social living. All that can be found is a gradual development of social attributes, suggesting . . . a substratum of social tendencies that extends the entire animal kingdom. From this substratum social life rises by the operation of different mechanisms and with various forms of expression until it reaches its present climax in vertebrates and insects" (1938, 274–75). Although Allee insisted on relaxing the strict boundaries between groups of subsocial animals, he still maintained a simpler hierarchy with certain groups attaining higher status with eusocial characteristics.

LEGITIMATING TERMS

Terms for social insects have been questioned and refined over time, as in the example of the "king bee" being renamed a "queen bee" once there was a realization that this "ruler of the monarchy" was a female. The case of the "white ants" presents a similar historical renaming. Initially, termites were thought to be a type of ant and therefore were called white ants; as this was discovered to be incorrect, the term slowly fell out of use. Although the terminology began to shift away from the label "white ants," references to "white ants" could be found as late as the 1940s, either citing the original term as a mistake or including the old term with the new one. In comparing bees to termites, Allee (1938, 265) made note of the terminology change: "The termites, miscalled white ants, belong to a relatively unspecialized insect order related to the cockroaches, and stand low in the evolutionary scale among the insects. They have, however, reached a high state of social development."

Unlike the term "white ants," some terms have remained contentious since the nineteenth century and at the same time continue to be used. As previously documented, the term "slave" was so strongly opposed by Huber that he presented an alternative term, "auxiliary." Forel argued that the new term was inadequate, whereas in this later account by Wheeler, the term is presented as the correct one. Wheeler described the *Formica fusca* as the "common black" ant found in the *Formica sanguinea* nest. "Thus the colony is mixed, and the black individuals, on account of their colour and provenience, have been called slaves. It is evident, however, that this term is inappropriate, for a slave is "a man who is the property of another, politically and socially at a lower level that the mass of the people, and performing compulsory labour" (Nieboer), and none of these distinctions applies to the *fusca* workers in the *sanguinea* nest. They might be more properly called 'auxiliaries'" (1928a, 285).

And yet, Wheeler himself generally employed the terms "slave-maker" and "slaves" in his work rather than "auxiliaries." Currently, this tends also to be the case with researchers who mention the mismatch of this analog in human culture and yet continue to use the concept of slavery to describe the behavior of social insects. Wheeler discussed the difficulty of changing terminology; his example concerned the use of the term "parasite" for a particular suborder (not a social insect) that differed from the behavior of parasites proper. He recounts the history of developing an alternative term, although none had been completely taken up in the dominant discourse. An abbreviated version of Wheeler's account offers a glimpse into the social process of renaming: "Attempts have been made to embody these distinctions in a term. . . . O. M. Reuter (1913, p. 53) introduced the term 'parasitoids' for these particular predators and I (1923) and more recently Root (1924) have employed this term. It is probable, however, that it will not be generally adopted and that 'parasite' will continue in vogue. If the distinction is clearly understood there can be no harm in such usage, but it should be noted that it has naturally created confusion in phylogenetic discussions" (1928a, 35). Wheeler seemed to acknowledge that despite incorrect usage a term can be "popular" and not seen as harmful, and therefore it will continue to be used in the literature. Similar to Kennedy's assertion about the use of anthropomorphic terms, Wheeler argued that there are consequences for using incorrect terminology, even for scientists who claim they are using terms metaphorically.

François Huber also expressed concern about terms but acquiesced to the current trend of usage. In anthropomorphically describing the actions of the hive in accepting a new queen, Huber claimed that falling back on certain terms was legitimate because the well-respected authority Réaumur used them. Despite an alternative tradition, the perpetuation of anthropomorphic terms and analogies can then, in part, be comprised of not only habit but also deference to tradition and authority. "I am sensible of the impropriety of these expressions, but M. de Réaumur in some respects authorises them. He does not scruple to say, that bees pay *attention, homage,* and *respect,* to their queen, and from his example the like expressions have escaped most authors that treat on bees" (1806, 131).

Oppositional discourse did exist throughout the nineteenth and early twentieth centuries. These ideas were a part of an "alternative tradition" that was in contention with many of the hierarchical concepts found in

Alternative Visions of Insect and (Human) Sociality

the dominant ideology of the time. These hierarchical concepts became embedded in the terminology and are foundational to ideas of sociality and organizational structure that still provoke debate today. Many of these current debates are extensions of the earlier alternative tradition and address the problems that remain from the legacy of race, class, and gender hierarchies that biased the development of concepts and theories surrounding sociality and social organization.

CURRENT ENTOMOLOGICAL DEBATES ON SOCIALITY

The alternative tradition continues as present-day entomologists question the dominant discourse involving the scale of sociality and the terms that are used to described social insects. It may be argued that as the methods for the study of insect behavior become more sophisticated, the use of particular anthropomorphic terms to describe the behavior becomes less significant. One can argue that, for contemporary entomologist, the terms have become unmoored from their referents; this is in fact the case made by many who use anthropomorphic terms. Within this line of reasoning, adhering to the terms becomes strictly a matter of convenience and continuity. However, as discussed throughout this book, scientists from the nineteenth century to the present day caution against the belief that they or their colleagues are immune to the effects of anthropomorphizing terms. A very recent statement made by Joan Herbers, an expert in the study of "slave-maker" ants, calls for other scientists to join her in relinquishing the use of slavery terminology. She argues that, "scientists must be open to the possibility that using racially loaded metaphors is inherently damaging to ourselves and to our work" (2006, B5).

More frequent in the contemporary literature are signs that opposition is building to the established definition and scales of sociality applied to insects. The hierarchical model used to classify social insects has been challenged on several levels. Costa and Fitzgerald (1996), in their article "Semantic Battles in a Conceptual War," present the history of the classification system for sociality, considering a combination of Michener's (1974) and Wilson's (1971) criteria to be the dominant paradigm (referred to as "M-W"). They then outline the "alternative conceptual frameworks" that began to develop starting in 1994 and conclude that most of these alternative frameworks simply call for a revision of the "M-W" dominant paradigm. Their belief is that the current dominant paradigm has caused a "conceptual neglect of structurally simple societies relative to the most complex societies" (1996, 289). In the vein of the earlier alternative tradition, they suggest that, "Students of social evolution must adopt a value-neutral conceptual context that is inclusive of all social organisms, and they must not focus unduly on traits exhibited by one subgroup" (Costa and Fitzgerald 1996, 289).

Although Costa and Fitzgerald question the categories of sociality, Costa in his recent book *The Other Insect Societies: Reconsidering the Insect Sociality Paradigm* (2006) acquiesces to the eusociality category, viewing it as capturing the unique sociality of ants, bees, wasps, and termites. What Costa does call for instead is dissolving the divisions between all other forms of sociality and to consider using the more inclusive criterion of communication to define sociality. He believes that this would result in the appreciation of the social complexity exhibited by other insects, such as caterpillars: "Beyond their utility as foils for comparative studies of insect societies, social caterpillars offer lessons in the marvelous complexity of insect behavior" (1997, 159). Costa's proposal to collapse the finer distinctions between levels of sociality other than eusociality is similar to Allee's earlier line of reasoning. The call for attention to nonsocial insect behavior continues to be an opposing discourse that decenters the emphasis on hierarchical behaviors for sociality criteria, while, at the same time, it leaves the term "eusociality" within the discourse.

Lacey and Sherman (2005, 573) question the definition of eusociality more directly. They have continued since 1995 to propose the use of a "eusociality continuum" based on a reproductive skew of alloparental species. This more general reproductive criterion includes cooperatively breeding species, noting that there may be a greater variation of reproductive skew based on ecological conditions. Taking this variation into account, species that had previously been excluded would be termed eusocial. Because their continuum model is more inclusive, they feel that the comparative value is actually greater. In an attempt to unify several proposed models, they combine the "eusociality continuum" with the alternative models of Crespi and Yanega (1995) and Wcislo (1997). In becoming exclusive to an extreme, the term "eusociality" loses some of its hierarchical status; in this opposing model, eusociality remains a workable category.

One of the defining features of sociality scales is the division of labor criterion, with eusociality becoming over time more strongly defined by a reproductive division of labor relying on caste. Deborah Gordon's early work began by questioning the dominant concept of caste for task allocation. She claimed that research emerging in the 1980s reflected findings that revealed a more flexible division of labor. She critiqued the idea of caste distribution as a rule, specifically the thesis presented in Oster and Wilson's 1978 book *Caste and Ecology in the Social Insects*. She faulted their hypothesis of caste along with their use of methodological technique, findings based on studies of small colonies within a laboratory setting (1989).

When Deborah Gordon began questioning the traditional understanding of the division of labor, she also began to try to understand what type of model would fit the task-switching behavior she herself had observed. Her alternative model of the behavior in an ant colony presented ants as governed by self-organization rather than participating in a hierarchical division of labor. In this model, there was no "queen" ordering "workers" to do particular tasks. Gordon likened the organization of ants to the functioning of a brain and utilized this metaphor to describe behavior and colony structure (1999). As she explained: "One class of models describing how simple, local rules generate global complexity are the neural networks. . . . It draws on the analogy between colonies and brains. In both systems, relatively simple units (ants or neurons), using local cues, can achieve complex, global behavior (1999, 143). This new model fits into the alternative tradition as it moves away from the previous ways of conceiving social insect societies hierarchically.

Some authors such as Marais and McCook had alluded to the presence of self-organization in social insect colonies, although they were not clear about its mechanism. Their observations at that time were used to critique the idea of caste or the capitalistic class-based analogies found in the dominant discourse. Deborah Gordon also questions this analogy, arguing that while it may appear "intuitively obvious" that specialized division of labor is more efficient, it may actually be less efficient than the work of individuals with the ability to perform multiple tasks (1989, 57). She believes the question can be asked of human industry as well, considering that the analogous term originated in this setting. Gordon points out that the idea of a highly specialized division of labor as the most efficient form of organization for humans has become controversial. She concludes that a specialized division of labor is definitely not the most efficient type of organization for insects and that therefore the use of this particular analogy was not a good fit (1989, 57).

Since Deborah Gordon's groundbreaking work, self-organizing models have become more accepted within the dominant discourse as way of describing both insect and human social organization. Interdisciplinary connections between sociology and entomology have resurfaced in the creation of these new models. Are self-organizing models the "way out" of hierarchical analogies? Do self-organizing models take into account the intersection of race, class, and gender as they are used to reconceptualize social behavior and organization? What lessons can be learned from the analysis of earlier dominant and alternative discourse of the nineteenth and early twentieth centuries? Careful attention must be paid to the history of embedded terms and assumptions in this literature. Hierarchical analogies and terms are not easily extricated from dominant ideas of sociality and social organization. Deconstructing an interlocking system of race, class, and gender hierarchies should take into consideration the varying implications of this socially constructed discourse.

Conclusion

During the nineteenth and early twentieth centuries, the disciplines of sociology and entomology co constructed analogies that compared social insect and human social organization. These analogies reinforced race, class, and gender hierarchies and became embedded in the concepts and terms used to describe sociality and social organization for both human and insect societies. The comparison between natural and social worlds created a legitimating loop that reinforced the credibility of these concepts and terms. Both sociology and entomology, as emerging disciplines, were aided by the effect of this legitimation.

The relationship between entomology and sociology in the nineteenth and early twentieth centuries included intellectual exchanges on major concepts, such as the division of labor. These conceptual exchanges were developed under larger paradigms of evolutionary theories in the nineteenth century and ecological theories in the early twentieth century. Comparisons between entomological and sociological phenomena came to be viewed as evidence for determining general laws of sociality and social organization. These general laws were framed as objective science, yet they were proposed and supported by social actors within a social context. The backgrounds of these actors and the sociopolitical influences of the time period factored into the creation of analogies between social insects and humans. The social construction of concepts and terms was influenced by the intersection of race, class, and gender, as well as other locations of time and place.

Understanding social construction is an important tool; however, critical theories are necessary to examine the issues of power found in the human/insect analogies. The critical science studies perspective used in this book not only exposes knowledge claims as socially constructed, but goes further to critique their hierarchical structure. The hierarchies that are created are naturalized not only as cultural objects, but also as reified power relations. Once naturalized, these hierarchies become embedded in the discourse and practices of society and must be excised with awareness of their potency.

An evaluation of power relations using critical discourse analysis revealed a systematic and interlocking set of hierarchies in the analogies between social insects and humans. Seen in this light, the earlier discourse takes on more significance and cannot be dismissed as random anthropomorphism unconnected to the creation of institutions or social practices. The analysis of these analogies present in the dominant discourse revealed an interlocking structure of race, class, and gender hierarchies. The influence of social evolutionary theories and of a Western colonial viewpoint was clearly present not only in the analogies, but also in the assumptions of the scale of sociality that ranks insects from the highest social group to the lowest. This scale used to define social insect societies was influenced by the same colonial worldview that informed the design of social evolutionary scales consisting of "higher" and "lower" human civilizations. The categories used in these scales were constructed by colonizers who viewed themselves as the highest in civilization and deemed those they colonized as the lowest. These hierarchical scales were also applied to other groups who were ranked by gender and class even though they shared the rank of the "civilized." The intersection of gender and class meant that women and the poor or working class were ranked lower on the scale of social evolution as well. The terminology used in the analogies reflected all of these hierarchies, with interlocking dichotomies such as "queen/worker" and "slave-maker/slave."

An analysis of the oppositional discourse illustrated the attempt to counter the race, class, and gender hierarchies found in the dominant discourse. Some of these oppositional analogies exemplified conscious feminist and socialist visions of the natural and social worlds. These visions reflected a more egalitarian interpretation of natural phenomena and used this to support proposed egalitarian social structures. A postcolonial viewpoint was the weakest theme in the oppositional discourse, so deeply was the colonial worldview embedded in the time period and the social location of those writing the discourse. In some work, however, traces remained of indigenous knowledge claims that would later be supplanted by Western redefinitions. These passages inadvertently contributed to an oppositional reading of natural phenomena that allowed for other perspectives.

Previously these historical data have been presented as mere anthropomorphism, a claim that denies or ignores the legitimating loop for social institutions and practices that occurs when natural and social

Conclusion

worlds are compared. Apolitical interpretations also downplay the important role of analogies in disseminating theoretical principles and concepts. These sociopolitical interactions are embedded in the naturalized concepts and terms still in use in the literature. The analysis presented in this book contributes to the growing debates concerning scales of sociality and calls for an infusion of feminist, postcolonial, and critical theory perspectives to inform new models and terminology. Because of the historical roots of the construction of these scales and the analogies used in the dominant discourse, the gendered, racialized, and classist themes are embedded in the basic terms and conceptions of social insects and filters through to newer models and the ongoing research.

The current interdisciplinary use of self-organization and of other naturalizing analogies that compare social insects, human organizations, and new technologies retains the hierarchical legacy of the earlier discourse. Current interpretation of social insect behavior and organizational structure is being coded into computer programs and used to reconfigure human organizations. Therefore, reconstructing ideas about insect eusociality has some consequence for overall paradigms of social organizational structure. Although self-organizing models appear to be a less hierarchical approach to understanding social organization, they may reify the social structure in other ways. Shouse (2002, 2357) reports on how the imagery of self-organization is being enacted on a global-technical scale that uses analogies of social insects to model computer systems for organizations:

> In a recent AT&T commercial, blue ants scurry through a maze and across a chasm in pursuit of an unknown goal. Only when the cartoon "camera" zooms out does the bigger picture come into view: The streams of ant traffic are forming the familiar lines of AT&T's logo. These fictitious bugs are a tongue-in-cheek illustration of how global order can emerge from what appears to be local chaos—a concept that computer programmers have been cribbing from insect behavior for years to make networks run more efficiently.

It is clear that while self-organizing systems accurately reflect certain aspects of complex group behavior, their comparison with a group may express reductionist assumptions about the individual "units" within that group, whether it is comprised of social insects or humans. Within a self-organizing system, the individual's action is viewed as less conscious and more a part of the larger global process. Kelly describes the actions of "hive mind" in superorganismic terms, analogizing these actions to "[a] typical day at the office." The self-organization of social insect colonies is viewed as a model that could parallel new decentralized work arrangements in human society.

On an even larger scale, the self-organizing model can be construed to explain free-market economic arrangements. "The marvel of 'hive mind' is that no one is in control, and yet an invisible hand governs, a hand that emerges from very dumb members" (Kelly 1994, 12). The lack of visible control, as Kelly plays off the metaphor, becomes a new version of Smith's invisible hand of the market. Self-organizing models are decentralized and nonhierarchical. However, it does not therefore follow that the individual takes on a more directly participatory role. Overall, the individual is seen as not having any idea of the more complex structure produced by their unconscious activity (Camazine et al. 2001, 268). Self-organizing models are a part of the larger emergent evolutionary theories. Emergence was embraced by earlier ecological conceptions of social insect models. Kelly (1994, 11) is aware of this as he traces this concept historically back to William Morton Wheeler: "Wheeler saw 'emergent properties' within the superorganism superseding the resident properties of the collective ants. Wheeler said that the superorganism of the hive 'emerges' from the mass of ordinary insect organisms."

Newer models do not drastically challenge the embedded hierarchies in the conceptions of sociality scales and social organization. Underlying assumptions that rest on race, class, and gender hierarchies remain unexplored. When founding categories and terminology are not questioned, they easily morph into other categories and terminology that are still based on framing the research in familiar ways. New anthropomorphic images emerge that align more closely with current institutions and practices, yet still maintain basic hierarchies. The newer terms and concepts are drawn from sociocultural sources, just as a close analysis of the dominant discourse of the nineteenth and early twentieth centuries revealed earlier terms and concepts concerning sociality and social organization.

Conclusion

Current research and new models occasionally create new terminology, but, in the main, they do not attempt to change the old terminology. These terms, as described in chapter 1, are almost all found intact within the current literature. Not addressing the need for changing the terminology leads to unnecessary and clumsy clarifications, as in the use of the terms "slave-makers" and "slaves" to describe a relationship that clearly does not fit that term; researchers may use quotation marks around the terms or other qualifications, but they then proceed to use the terms throughout. Many entomologists believe that the social meanings of these terms no longer have any bearing on their usage, although others in the discipline argue that the embedded social meanings remain significant for entomologists and those that are exposed to their research.

Some terms with a history of being questioned also have a pattern of being used despite the problematic fit. Two terms in this category are "queen" and "slave." Early on, some entomologists felt that the term "queen" might be misleading. This term is still in use among current researchers; although in self-organizing models the role has been further deemphasized and has lost "queenlike" status. And what is most striking concerning the term "slave" is that, despite that an alternative—and possibly more accurate—term was created early in the nineteenth century to describe this behavior, the terms surrounding slavery still abound in the dominant discourse. It would seem that the term "slave" is seen as being better suited for analogies to human society; there is no clear analog to "auxiliaries" that would reinforce a hierarchical social structure in the way that slavery does. To argue that current use of the term "slavery" is insignificant because slavery has been legally abolished in Western countries would be to miss the point. The acceptance of the terms that describe slavery and the extreme hierarchy it implies leave this as a naturalized institution.

It is necessary to critically assess the continued use of these terms in the new models and in everyday usage within entomology. Some terms such as "worker" have no history of being questioned. A relationship of hierarchy for workers is depicted when the term "queen" is retained, and this relational factor may be thought to provide a reason for retaining the term despite a long history of challenges regarding its appropriate applicability. This is true of the terms "slave-maker" and "slave" as well.

Dynamics of race, class, and gender are not isolated anthropomorphic terms but interlock into a hierarchical system; naturalizing this system is dependent on the interlocked terms. Specifically, some of these new models do not resolve any of the problems of the embeddedness of assumptions regarding class, race, and gender. In part, this is due to leaving the original terminology in place, which then contributes to perpetuating the use of analogies involving race, class, and gender hierarchies. The larger-domain assumptions behind the terminology also seem to survive, such as the legacy of a colonial worldview. If new visions of sociality are to be developed, it is imperative to employ postcolonial theory as a tool for understanding the construction of this sociality scale. The debates in the field of entomology over the associated problems of the sociality scale attest to the lingering assumptions that hamper research. Not only examining the standpoint of models and metaphors but also acknowledging that other worldviews create different models, metaphors, and terms would be extremely significant for reducing the Western bias that remains in the assessment of sociality and social organization. Sociohistorical information is instructive for this purpose, as is the research coming from current ethnoentomology. Postcolonial science studies seek to decenter the Western claim to objective scientific knowledge without seeking to replace this with an essentialized indigenous knowledge.

In line with this need for a critique of Western bias, are the new self-organizing models a particularly Western discursive tool? Are they a part of a "technical turn" that sees nature and social not as a machine in the more traditional Cartesian sense but in the newer, computer-networking sense of the word? Is this any more accurate? And further, does this model eliminate race, class, and gender hierarchies from the underlying assumptions or analogies? Although self-organizing models originally were touted as moving away from hierarchical understandings of organization, the real difference has been the lack of visible, external hierarchies. Underscoring this, current propositions in the literature include discussions of self-organizing dominance hierarchies and internal hierarchies of the individual based on the simple rules of stimulus-response (Beshers, Robinson, and Mittenthal 1999).

As self-organizing systems deemphasize the individual, the computer simulation programs and artificial intelligence that utilize these self-

Conclusion

organizing models are also viewed as neutral sources of evidence about natural and social systems. However, computer programs and AI are constructed by humans and subject to the cultural biases of these humans. In attempting to create feminist AI projects, Adam (2001, 349) charts the manner in which artificial intelligence projects are "inscribed with a view from mainstream epistemology." She further maintains that the subject for AI fits the "masculine rationalist ideal." Whereas essentializing feminist models as found in the earlier alternative discourse would not be helpful, awareness of gender is crucial in the development of new models.

Within feminist critiques of primatology, areas that had once been considered neutral were found to be informed by deeply embedded gender biases. Given the history of the analogies between humans and social insects, the implication of gender bias remains a part of the legacy of the discourse. Statements in current research articles that describe the queen still make use of gendered metaphors such as: "Queens may adopt strategies that allow them to dominate rivals, either by gaining an advantage in fights among the queens or by increasing their attractiveness to workers" (Balas 2004, 77). Even more, overall conceptions of the reproduction strategies and organization of social insects still involve deep gender stereotypes. Some discussion of gender must occur with the development of new models and conceptions of sociality and social organization. This discussion must intersect with other hierarchies, as critical feminist science studies has done in the field of primate studies.

Sociality and social organization are being redefined, and these reconfigurations affect both the interpretation of insect and human individuals and societies. Descriptions of human and insect societies have a history of shared concepts and theories that were tied to hierarchical analogies. Newer models continue to use these analogies, expanding them to include the computer technologies that have become so significant in current interpretation of interactions and structures. As debates within the field of entomology attempt to grapple with the legacy of the founding conceptions and theories, and to replace these with new paradigms, an understanding of how a legitimating loop was formed between the natural and the social will be necessary for a truly critical paradigm change. With this sociological analysis of the analogies between social insects and humans that were common in the dominant discourse of the nineteenth and early twentieth centuries, I hope to bring attention to how deeply embedded are the race, class, and gender hierarchies in these mutual understandings of sociality and social organization. This has implications for both the understanding of the natural world as well as the structuring of the social world. I believe that this embeddedness calls for a critical post-Kuhnian, feminist, and postcolonial science studies approach to the current debates on sociality and creation of new models.

Ideas about sociality and social organization are not abstract propositions. They have consequences for everyday practices. While the behavior of social insects does not become more or less hierarchical with the trends in human interpretation of their behavior, these interpretations do provide a naturalizing tool to reinforce human institutions and practices. In this regard, these interpretations are significant and should be addressed by sociology as well as entomology. In applying social conceptions to the organization and behavior of insects and then reflecting these back as natural, a legitimating loop occurs that reinforces particular social structures. Denaturalizing these social theories and concepts must be done in many arenas, and the interactions between sociology and entomology must be one of these focal points.

References

Abraham, Itty. 2000. "Postcolonial Science, Big Science, and Landscape." In *Doing Science + Culture*, edited by Roddey Reid and Sharon Traweek, 49–70. New York: Routledge.
Adam, Alison. 2001. "Feminist AI Projects and Cyberfutures." In *Women, Science and Technology: A Reader in Feminist Science Studies*, edited by Mary Wyer, Mary Barbercheck, Donna Geisman, Hatice Örün Öztürk, and Marta Wayne, 332–54. New York: Routledge.
Alger, Janet M., and Steven F. Alger. 1997. "Beyond Mead: Symbolic Interaction between Humans and Felines." *Society and Animals* 5 (1):65–81.
Alic, Margaret. 1986. *Hypatia's Heritage: A History of Women in Science from Antiquity through the Nineteenth Century*. Boston: Beacon Press.
Allee, W. C. 1931a. *Animal Aggregations: A Study in General Sociology*. Chicago: University of Chicago Press.
———. 1931b. "Co-operation among Animals." *American Journal of Sociology* 37:386–98.
———. 1938. *The Social Life of Animals*. New York: Norton.
Allen, Danielle. 2004. "Burning the Fable of the Bees: The Incendiary Authority of Nature." In *The Moral Authority of Nature*, edited by Lorraine Daston and Fernando Vidal, 74–99. Chicago: University of Chicago Press.
Allen, Grant. 1904. "Personal Reminiscences of Herbert Spencer (1894)." *Forum* 35: 610–68.
Ancarani, Vittorio. 1995. "Globalizing the World: Science and Technology in International Relations." In *Handbook of Science and Technology Studies*, edited by Sheila Jasanoff, Gerald E. Markle, James C. Petersen, and Trevor Pinch, 652–70. Thousand Oaks, CA: Sage.
Anonymous. 1837. *The Natural History of Insects*. Vol. 1. New York: Harper.
Ardrey, Robert. 1969. Introduction to *The Soul of the Ape*, by Eugène Marais, 1–55. New York: Atheneum.
Arluke, Arnold, and Clinton R. Sanders. 1996. *Regarding Animals*. Philadelphia: Temple University Press.
Asquith, Pamela J. 1996. "Japanese Science and Western Hegemonies: Primatology and the Limits Set to Questions." In *Naked Science: Anthropological Inquiry into Boundaries, Power, and Knowledge*, edited by Laura Nader, 239–58. New York: Routledge.
Balas, M. T. 2004. "Conditions Favoring Queen Execution in Young Social Insect Colonies." *Insectes Sociaux* 52:77–83.

Banks, Edwin M. 1985. "Warder Clyde Allee and the Chicago School of Animal Behavior." *Journal of the History of the Behavioral Sciences* 21 (4):345–53.

Barnes, Barry, and Steven Shapin, eds. 1979. *Natural Order: Historical Studies of Scientific Culture*. London: Sage.

Barnes, Jeffery K. 1985. "Insects in the New Nation: A Cultural Context for the Emergence of American Entomology." *Bulletin of the Entomological Society of America* 31 (1):21–30.

Batra, S. W. T. 1966. "Nests and Social Behavior of Halictine Bees of India (Hymenoptera: Halictidae)." *Indian Journal of Entomology* 28:375–93.

Beck, Alan M. 1996. "The Common Qualities of Man and Beast." *Chronicle of Higher Education* 42 (May 17, 1996):B3.

Becker, Howard, and Harry Elmer Barnes. 1961. *Social Thought from Lore to Science*. New York: Dover.

Becker, Lydia E. [1868] 1996. "Is There Any Specific Distinction between Male and Female Intellect?" In *Gender and Science: Late Nineteenth-Century Debates on the Female Mind and Body*, edited by Katharina Rowold, 15–22. Bristol, England: Thoemmes Press.

Beebe, William. 1921. *Edge of the Jungle*. New York: Holt.

Beier, Max. 1973. "The Early Naturalists and Anatomists during the Renaissance and Seventeenth Century." In *History of Entomology*, edited by Ray F. Smith, Thomas E. Mittler, and Carroll N. Smith, 81–94. Palo Alto, CA: Annual Reviews.

Belt, Thomas. 1911. *The Naturalist in Nicaragua*. London: Dent.

Bentley, Jeffery W., and Gonzalo Rodríguez. 2001. "Honduran Folk Entomology." *Current Anthropology* 42 (2):1–16.

Berlin, Brent. 1992. *Ethnobiological Classification: Principles of Categorization of Plants and Animals in Traditional Societies*. Princeton: Princeton University Press.

Beshers, Samuel N., Gene E. Robinson, and Jay E. Mittenthal. 1999. "Response Thresholds and Division of Labor in Insect Colonies." In *Information Processing in Social Insects*, edited by Claire Detrain, Jean-Louis Deneubourg, and Jacques Pasteels, 115–39. Basel, Switzerland: Birkhäuser Verlag.

Birch, David. 1989. *Language, Literature and Critical Practice: Ways of Analyzing Text*. London: Routledge.

Blu Buhs, Joshua. 2000. "Building on Bedrock: William Steel Creighton and the Reformation of Ant Systematics, 1925–1970." *Journal of the History of Biology* 33:27–70.

Bogardus, Emory. [1918] 1920. *Essentials of Social Psychology*. 2nd ed. Los Angeles: University of Southern California Press.

Boswell, Graeme P., Nigel R. Franks, and Nicholas F. Britton. 2001. "Arms Races

and the Evolution of Big Fierce Societies." *Proceedings: Biological Sciences B* 268 (1477):1723–30.
Bowker, Geoffrey C., and Susan Leigh Star. 1999. *Sorting Things Out: Classification and Its Consequences.* Cambridge: MIT Press.
Burian, Richard M. 1996. "Some Epistemological Reflections on Polistes as a Model Organism." In *Natural History and Evolution of an Animal Society: The Paper Wasp Case,* edited by Stephano Turillazzi and Mary Jane West-Eberhard, 318–37. Oxford: Oxford University Press.
Burke, Peter. 1997. "Fable of the Bees: A Case-Study in Views of Nature and Society." In *Nature and Society in Historical Context,* edited by Mikulás Teich, Roy Porter, and Bo Gustafsson, 112–23. Cambridge: Cambridge University Press.
Burton, Frances. 1994. "In the Footsteps of Heraclitus." In *Strength in Diversity: A Reader in Physical Anthropology,* edited by A. Herring and L. Chan, 77–102. Toronto: Canadian Scholars' Press.
Camazine, Scott, Jean-Louis Deneubourg, Nigel R. Franks, James Sneyd, Guy Theraulaz, and Eric Banabeau. 2001. *Self-Organization in Biological Systems.* Princeton: Princeton University Press.
Carneiro, Robert L., ed. 1967. *The Evolution of Society: Selections from Herbert Spencer's Principles of Sociology.* Chicago: University of Chicago Press.
Chauvin, Remy. [1963] 1968. *Animal Societies: From the Bee to the Gorilla.* New York: Hill and Wang.
Cherry, Lynne. 1988. "Achieving Scientific Accuracy in Children's Nature Books." *Science Books & Film* 34 (8): 9–15.
Clark, J. F. M. 1997. "'The Ants Were Duly Visited': Making Sense of John Lubbock, Scientific Naturalism and the Senses of Social Insects." *British Journal for the History of Science* 30:151–76.
———. 1998. "'The Complete Biography of Every Animal': Ants, Bees and Humanity in Nineteenth-Century England." *Studies in History and Philosophy of Biology and Biomedical Sciences* 29 (2): 249–67.
Clark, Linda L. 1981. "Social Darwinism in France." *Journal of Modern History* 53 (1): D1025–44.
Clutton-Brock, Juliet. 1999. "Aristotle, the Scale of Nature, and Modern Attitudes to Animals." In *Humans and Other Animals,* edited by Arien Mack, foreword by Marc Bekoff, 5–24. Columbus: Ohio State University Press.
Cohen, I. B. 1994. *Interactions: Some Contacts between the Natural Sciences and the Social Sciences.* Cambridge: MIT Press.
Collias, Nicholas E. 1991. "The Role of American Zoologists and Behavioural Ecologists in the Development of Animal Sociology, 1934–1964." *Animal Behaviour* 41:613–31.

Comstock, Anna. B. 1905. *How to Keep Bees*. New York: Doubleday, Page.
Costa, James T. 1997. "Caterpillars as Social Insects." *American Scientist* 85:150–59.
———. 2006. *The Other Insect Societies: Reconsidering the Insect Sociality Paradigm*. Cambridge: Harvard University Press.
Costa, James T., and Terrence D. Fitzgerald. 1996. "Semantic Battles in a Conceptual War." *Trends in Ecology and Evolution*. 11 (7):285–89.
———. 2005. "Social Terminology Revisited: Where Are We Ten Years Later?" *Annales of Zoologici Fennici* 42:559–64.
Cox, Oliver C. 1942. "The Modern Caste School of Race Relations." *Social Forces* 21:218–26.
Crane, Eva. 1975. "The World's Beekeeping—Past and Present." In *Facts from the Hive and the Honeybee*, edited by Dadant and Sons, 1–18. Hamilton, IL: Dadant.
Crespi, Bernard J., and Douglas Yanega. 1995. "The Definition of Eusociality." *Behavioral Ecology* 6 (1):109.
Crist, Eileen. 1999. *Images of Animals: Anthropomorphism and Animal Mind*. Philadelphia: Temple University Press.
———. 2004. "Can an Insect Speak? The Case of the Honeybee Dance Language." *Social Studies of Science* 34 (1):7–43.
Cromie, William J. 2005. "Taking a Look at How Ant (And Human) Societies Might Grow." Harvard University Gazette, September 29, 2005. www.news.harvard.edu/gazette/2005/09.29/13-colonies.html.
Cronin, Helena. 1991. *The Ant and the Peacock: Altruism and Sexual Selection from Darwin to Today*. Cambridge: Press Syndicate of the University of Cambridge.
Curley, Edwin. 1885. "Bees and Other Hoarding Insects: Their Specialization into Females, Males and Workers." *Entomologica Americana* 1 (4):61–72.
Dalke, Kate. 2003. "The Queen Bee's Allure." *Genome News Network*. www.genomenewsnetwork.org/articles/04_03/bee.shtml.
Darwin, Charles. [1859] 1998. *The Origin of Species*. New York: Modern Library.
———. 1874. *The Descent of Man*. New York: Collier.
Davis, Natalie Zemon. 1995. *Women on the Margins: Three Seventeenth-Century Lives*. Cambridge: Harvard University Press.
Dean, John. 1979. "'The Controversy over Classification': A Case Study from the History of Botany." In *Natural Order: Historical Studies of Scientific Culture*, edited by Barry Barnes and Steven Shapin, 211–28. London: Sage.
de Waal, Frans. 1989. *Peacemaking among Primates*. Cambridge: Harvard University Press.
Dickens, Peter. 2000. *Social Darwinism: Linking Evolutionary Thought to Social Theory*. Philadelphia: Open University Press.

Douglas, Mary. 1986. *How Institutions Think*. Syracuse, NY: Syracuse University Press.
Dubow, Saul, ed. 2000. *Science and Society in Southern Africa*. New York: St. Martin's Press.
Duit, R. 1991. "On the Role of Analogies and Metaphors in Learning Science." *Science Education* 5 (6):649–72.
Durkheim, Emile. [1893] 1933. *The Division of Labor*. New York: Macmillan.
Eimer, G. H. Theodor. 1890. *Organic Evolution*. London: Macmillan.
Ellen, Roy. 1993. *The Cultural Relations of Classification: An Analysis of Nuaulu Animal Categories from Central Seram*. Cambridge: Cambridge University Press.
Ellwood, Charles A. 1901. "The Theory of Imitation in Social Psychology." *American Journal of Sociology* 6:721–41.
———. 1907. "Sociology: Its Problems and Relations." *American Journal of Sociology* 13 (3):300–348.
———. 1910. *Sociology and Modern Social Problems*. New York: American Book Company.
Emerson, Alfred E. 1942. "Basic Comparisons of Human and Insect Societies." In *Biological Symposia VIII, Levels of Integration in Biological and Social Science*, edited by Robert Redfield, 163–76. Lancaster, PA: Jaques Cattell Press.
———. 1943. "Ecology, Evolution and Society." *American Naturalist* 77 (769):97–118.
Engels, Frederick. [1883] 1940. *Dialectics of Nature*. New York: International.
Escobar, Arturo. 1992. "Culture, Economics, and Politics in Latin American Social Movements Theory and Research." In *The Making of Social Movements in Latin America: Identity, Strategy, and Democracy*, edited by Arturo Escobar and Sonia E. Alvarez, 62–85. Boulder, CO: Westview Press.
Espinas, Alfred. 1878. *Des sociétés animales*. Translated passages by Isabelle Roughol. Paris: Bailliére.
Evans, Howard E. 1984. *Insect Biology: A Textbook of Entomology*. London: Addison-Wesley.
Evans, T. A. 2006. "Foraging and Building in Subterranean Termites: Task Switchers or Reserve Labourers?" *Insectes Sociaux* 53 (1):56–64.
Evrard, Eugene. 1923. *The Mystery of the Hive*. Translated by Bernard Miall. New York: Dodd, Mead.
Fabre, J. H. 1912. *Social Life in the Insect World*. Translated by Bernard Miall. London: Fisher Unwin.
———. 1938. *Marvels in the Insect World*. Edited, annotated, and translated by Percy F. Bicknell. New York: Appleton-Century.
Fairclough, Norman. 1989. *Language and Power*. London: Longman.
———. 1995. *Critical Discourse Analysis: The Critical Study of Language*. London: Longman.

Fichman, Martin. 1997. "Biology and Politics: Defining the Boundaries." In *Victorian Science in Context*, edited by Bernard Lightman, 94–118. Chicago: University of Chicago Press.

Forel, Auguste. 1904. *Ants and Some Other Insects*. Chicago: Open Court.

———. 1928. *The Social World of the Ants Compared with That of Man*. London: Putnam's.

Foster, William A. 2002. "Soldier Aphids Go Cuckoo." *Trends in Ecology and Evolution* 17 (5):199–200.

Foucault, Michel. 1973. *The Order of Things: An Archaeology of the Human Sciences*. New York: Vintage.

Franks, Nigel R. 2001. "Evolution of Mass Transit Systems in Ants: A Tale of Two Societies." In *Insect Movement: Mechanisms and Consequences*. Proceedings of the 20th Symposium of the Royal Entomological Society, 281–98. CAB International.

Freeman, Scott, and Jon C. Herron. 2004. *Evolutionary Analysis*. 3rd ed. Upper Saddle River, NJ: Pearson Prentice Hall.

Gailey, Christine Ward. 1996. "Politics, Colonialism, and the Mutable Color of Pacific Islanders." In *Race and Other Misadventures: Essays in Honor of Ashley Montagu in His Ninetieth Year*, edited by Larry T. Reynolds and Leonard Lieberman, 36–49. Dix Hills, NY: General Hall.

Gardner A., and S. A. West. 2004. "Spite and the Scale of Competition." *Journal of Evolutionary Biology* 17:1195–203.

Gates, Barbara T. 1997. "Ordering Nature: Revisioning Victorian Science Culture." In *Victorian Science in Context*, edited by Bernard Lightman, 179–86. Chicago: University of Chicago Press.

Gerard R. W. 1942. "Higher Levels of Integration." In *Biological Symposia VIII, Levels of Integration in Biological and Social Science*, edited by Robert Redfield, 67–87. Lancaster, PA: Jaques Cattell Press.

Giddings, Franklin Henry. 1896. *Principles of Sociology*. New York: Macmillan.

———. 1901. *Inductive Sociology: A Syllabus or Methods, Analysis and Classifications, and Provisionally Formulated Laws*. New York: Macmillan.

———. 1932 (posthumous). *Civilization and Society: An Account of the Development and Behavior of Human Society*. Arranged and edited by Howard W. Odum. New York: Holt.

Gieryn, Thomas. 1983. "Boundary Work and the Demarcation of Science from Non-Science: Strains and Interests in Professional Ideologies of Scientists." *American Sociological Review* 48:781–95.

Gilbert, Nigel, and Michael Mulkay. 1984. *Opening Pandora's Box: A Sociological Analysis of Scientists' Discourse*. Cambridge: Cambridge University Press.

Gilman, Charlotte Perkins. 1911. *The Man-Made World; or, Our Androcentric Culture*. New York: Charlton.

Gilman, Charlotte Stetman. 1898. *Women and Economics: A Study of the Economic Relation between Men and Women as a Factor in Social Evolution.* Boston: Small, Maynard.

Gordon, Deborah. 1989. "Caste and Change in Social Insects." *Oxford Surveys in Evolutionary Biology,* vol. 6, edited by Paul H. Harvey and Linda Partridge. Oxford: Oxford University Press.

———. 1999. *Ants at Work.* New York: Free Press.

Gornick, Vivian. 1983. *Women in Science: Portraits from a World in Transition.* New York: Simon and Schuster.

Gould, Stephen Jay. 1981. *The Mismeasure of Man.* New York: Norton.

———. 1997. "Kropotkin Was No Crackpot." *Natural History* 106:12–21.

Graham, Loren R. 1993. *Science in Russia and the Soviet Union: A Short History.* New York: Cambridge University Press.

Greene, John C. 1981. *Science, Ideology, and World View: Essays in the History of Evolutionary Ideas.* Berkeley and Los Angeles: University of California Press.

Gross, Matthias. 2002. "When Ecology and Sociology Meet: The Contributions of Edward A. Ross." *Journal of the History of the Behavioral Sciences* 38 (1):27–42.

Gross, Neil, and Robert Alan Jones, eds. and trans. 2004. *From Durkheim's Philosophy Lectures: Notes from the Lycée de Sens Course, 1883–1884.* Cambridge: Cambridge University Press.

Gurung, Astrid Björnsen. 2003. "Insects—A Mistake in God's Creation? Tharu Farmers' Perception and Knowledge of Insects: A Case Study of Gobardiha Village Development Committee, Dang-Deukhuri, Nepal." *Agriculture and Human Values* 20:337–70.

Hamilton, William D. 1964. "The Genetical Evolution of Social Behaviour I and II." *Journal of Theoretical Biology* 7:1–16, 17–52.

Haraway, Donna. 1989. *Primate Visions: Gender, Race and Nature in the World of Modern Science.* New York: Routledge.

———. 1991. *Simians, Cyborgs, and Women: The Reinvention of Nature.* New York: Routledge.

Harding, Sandra. 1998. *Is Science Multicultural?: Postcolonialisms, Feminisms, and Epistemologies.* Bloomington: Indiana University Press.

Harries, Patrick. 2000. "Field Sciences in Scientific Fields: Entomology, Botany and the Early Ethnographic Monograph in the Work of H.-A. Junod." In *Science and Society in Southern Africa,* edited by Saul Dubow, 11–41. New York: St. Martins Press.

Harvey, D. 1996. *Justice, Nature and the Geography of Difference.* Oxford: Blackwell.

Harwood, Jonathon. 1987. "National Styles in Science: Genetics in Germany and the United States between the World Wars." *Isis* 78 (3):390–414.

Heinze, J., S. Cremer, N. Eckl, and A. Schrempf. 2006. "Stealthy Invaders: The Biology of *Cardiocondyla* Tramp Ants." *Insectes Sociaux* 53 (1):1–7.

Herbers, Joan M. 2006. "The Loaded Language of Science." *Chronicle of Higher Education* 52 (29): B5, 1p, 1c.

Hermann, Henry R. 1979. *Social Insects*. Vol. 1. New York: Academic Press.

Herzenberg, Caroline. 1986. *Women Scientists from Antiquity to the Present*. West Cornwall, CT: Locust Hill Press.

Hess, David J. 1995. *Science and Technology in a Multicultural World: The Cultural Politics of Facts and Artifacts*. New York: Columbia University Press.

Hesse, Mary B. 1966. *Models and Analogies in Science*. Notre Dame: University of Notre Dame Press.

Hilts, Victor L. 1994. "Towards the Social Organism: Herbert Spencer and William B. Carpenter on the Analogical Method." In *The Natural Sciences and the Social Sciences*, edited by I. B. Cohen, 275–303. Dordrecht: Kluwer Academic Press.

Hofstadter, Richard. 1955. *Social Darwinism in American Thought*. Boston: Beacon Press.

Hölldobler, Bert, and Edward O. Wilson. 1990. *The Ants*. Cambridge: Belknap Press of Harvard University Press.

Huber, François. 1806. *New Observations on the Natural History of Bees*. Translated from the original. Edinburgh: Printed for John Anderson and sold by Longman, Hurst, Rees and Orms: London.

Hull, David. 1992. "Biological Species: An Inductivist's Nightmare." In *How Classification Works: Nelson Goodman among the Social Sciences*, edited by Mary Douglas and David Hull, 42–68. Edinburgh: Edinburgh University Press.

Huxley, Julian. 1930. *Ants*. New York: Jonathon Cape and Harrison Smith.

Huxley, Thomas. 1894. "Prolegomena." In *Evolution and Ethics and Other Essays*, 1896, 1–45. New York: Appleton.

Irvine, Leslie. 2004. "A Model of Animal Selfhood: Expanding Interactionist Possibilities." *Symbolic Interaction* 27 (1):3–21.

Jones, Greta. 1980. *Social Darwinism and English Thought: The Interaction between Biological and Social Theory*. Sussex: Harvester Press.

Keller, Evelyn Fox. 1985. *Reflections on Gender and Science*. New Haven: Yale University Press.

———. 1995. "The Origin, History, and Politics of the Subject Called 'Gender and Science': A First Person Account." In *Handbook of Science and Technology Studies*, edited by Sheila Jasanoff, Gerald E. Markle, James C. Petersen, Trevor Pinch, 80–94. Thousand Oaks, CA: Sage.

Kellogg, Vernon. 1908. *American Insects*. New York: Holt.

Kelly, Kevin. 1994. *Out of Control: The Rise of Neo-Biological Civilization.* Reading, MA: Addison-Wesley.

Kennedy, John S. 1992. *The New Anthropomorphism.* Cambridge: University of Cambridge Press.

Kingsland, Sharon E. 2005. *The Evolution of American Ecology, 1890–2000.* Baltimore: Johns Hopkins University Press.

Klamer, Arjo, and Thomas C. Leonard. 1994. "So What's an Economic Metaphor?" In *Natural Images in Economic Thought: "Markets Read in Tooth and Claw,"* edited by Philip Mirowski, 20–54. Cambridge: Cambridge University Press.

Knorr-Cetina, Karin D. 1981. *The Manufacture of Knowledge: An Essay on the Constructivist and Contextual Nature of Science.* Oxford: Pergamon Press.

Kropotkin, Peter. 1902. *Mutual Aid.* New York: McClure, Phillips.

Kuhn, Thomas S. 1962. *The Structure of Scientific Revolutions.* Chicago: University of Chicago Press.

Lacey, Eileen A., and Paul W. Sherman. 2005. "Redefining Eusociality: Concepts, Goals and Levels of Analysis." *Annales of Zoologici Fennici* 42:573–77.

Lakoff, George. 1987. *Women, Fire, and Dangerous Things: What Categories Reveal about the Mind.* Chicago: University of Chicago Press.

Lakoff, George, and Mark Johnson. 1980. *Metaphors We Live By.* Chicago: University of Chicago Press.

Latour, Bruno, and Steve Woolgar 1979. *Laboratory Life: The Social Construction of Scientific Facts.* Beverly Hills, CA: Sage.

Latter, O. H. 1913. *Bees and Wasps.* Cambridge: Cambridge University Press.

Levine, Daniel. 1995. "The Organism Metaphor in Sociology." *Social Research.* 62 (2):239–65.

Lewontin, R. C. 1991. *Biology as Ideology.* New York: Harper Perennial.

Limoges, Camille. 1994. "Milne-Edwards, Darwin, Durkheim and the Division of Labour: A Case Study in Reciprocal Conceptual Exchanges between the Social and the Natural Sciences." In *The Natural Sciences and the Social Sciences,* edited by I. B. Cohen, 317–43. Dordrecht: Kluwer Academic Press.

Lovejoy, Arthur O. 1936. *The Great Chain of Being: A Study of the History of an Idea.* Cambridge: Harvard University Press.

Lubbock, Sir John. 1882. *Ants, Bees and Wasps.* London: Kegan, Paul, Trench.

Lutz, Frank. 1941. "The Truth about Termites." *Natural History* 48 (2):113–15.

Lynch, Michael. 1982. "Technical Work and Critical Inquiry: Investigations into a Scientific Laboratory." *Social Studies of Science* 12:499–533.

Maeterlinck, Maurice. 1901. *The Life of the Bee.* New York: Dodd, Mead.

Marais, Eugène. [1937] 1973. *The Soul of the White Ant.* Middlesex, England: Penguin.

Margulis, Lynn. 1998. *Symbiotic Planet: A New Look at Evolution.* New York: Basic.

Marks, Jonathan. 2002. *What it Means to Be 98% Chimpanzee: Apes, People and Their Genes.* Berkeley and Los Angeles: University of California Press.

Martin, Emily. 1991. "The Egg and the Sperm: How Science Has Constructed a Romance Based on Stereotypical Male-Female Roles." *Signs* 16:485–501.

Martindale, Don. 1960. *The Nature and Types of Sociological Theory.* Boston: Houghton Mifflin.

Marx, Karl, and Frederick Engels. [1862] 1986. *Collected Works of Marx and Engels.* Vol. 41, *Correspondence 1860–1864.* New York: International.

Maus, Heinz. 1962. *A Short History of Sociology.* London: Routledge and Kegan Paul.

Mayberry, Maralee, Banu Subramaniam, and Lisa H. Weasel. 2001. *Feminist Science Studies: A New Generation.* New York: Routledge.

Mayr, Ernst. 1977. "Darwin and Natural Selection." *American Scientist* 65 (3):321–27.

McCook, Henry Christopher. 1909. *Ant Communities and How They Are Governed: A Study in Natural Civics.* New York: Harper.

McDougall, William. 1920. *The Group Mind.* New York: Putnam's.

Mead, George Herbert. 1934. *Mind, Self and Society.* Chicago: University of Chicago Press.

Meek, Ronald L, ed. 1953. *Marx and Engels on Malthus: Selections from the Writings of Marx and Engels Dealing with the Theories of Thomas Robert Malthus.* London: Lawrence and Wishart.

Merrick, Jeffrey. 1988. "Royal Bees: The Gender Politics of the Beehive in Early Modern Europe." *Studies in Eighteenth-Century Culture* 18:7–37.

Michener, Charles. 1974. *The Social Behavior of Bees.* Cambridge: Harvard University Press.

Michener, Charles, and Mary H. Michener. 1951. *American Social Insects.* New York: Van Nostrand.

Mikulinsky, S. R. 1981. "The Historical Conditions and Features of the Development of Natural Science in Russia in the First Half of the 19th Century." In *Epistemological and Social Problems of the Sciences in the Early Nineteenth Century*, edited by H. N. Jahnke and M. Otte, 91–107. Dordrecht, Holland: Reidel.

Milius, Susan. 2000. "Slavemaker Ants: Misunderstood Farmers?—Behavior Examined." *Science News*, August 19, 2000.

Mirowski, Philip, ed. 1994. *Natural Images in Economic Thought: "Markets Read in Tooth and Claw."* Cambridge: University Cambridge Press.

Mitchell, Sandra. D. 2002. "Integrative Pluralism." *Biology and Philosophy* 17:55–70.

Mitman, Gregg. 1988. "From the Population to Society: The Cooperative Metaphors of W. C. Allee and A. E. Emerson." *Journal of the History of Biology* 21 (2):173–94.

———. 1992. *The State of Nature: Ecology, Community and American Social Thought, 1900–1950*. Chicago: University of Chicago Press.

Morgan, C. Lloyd. 1895. *Animal Life and Intelligence*. Boston: Ginn.

———. 1908. *Animal Behavior*. London: Arnold.

Mukerjee, Radhakamal. 1926. *Regional Sociology*. New York: Century.

———. 1942. *Social Ecology*. London: Longmans, Green.

Mulkay, Michael. 1985. *The Word and the World: Explorations in the Form of Sociological Analysis*. London: Allen and Unwin.

Nandy, Ashis. 1988. *Science, Hegemony and Violence: A Requiem for Modernity*. Delhi: Oxford University Press.

Novellino, Dario. 2000. "The Relevance of Myths and Worldviews in Pälawan Classification, Perceptions, and Management of Honey Bees." In *Ethnobiology and Biocultural Diversity: Proceedings of the Seventh International Congress of Ethnobiology*, edited by John R. Stepp, Felice S. Wyndham, and Rebecca K. Zarger, with assistance from Mika Cohen and Sarah Lee, 189–206. Athens: University of Georgia Press.

Ogburn, William Fielding, and Alexander Goldenweiser, eds. 1927. *The Social Sciences and Their Interrelations*. Boston: Houghton Mifflin.

Olson, Richard. 1971. *Science as Metaphor: The Historical Role of Scientific Theories in Forming Western Culture*. Belmont, CA: Wadsworth.

Oster, George F., and Wilson, E. O. 1978. *Caste and Social Ecology in the Social Insects*. Princeton: Princeton University Press.

Owen, Denis F. 1990. "The Language of Attack and Defense." *Oikos* 57 (1):133–35.

Park, Robert Ezra. 1936. "Human Ecology." *American Journal of Sociology* 42 (1):1–15.

Parker, George Howard. 1938. "Biographical Memoir of William Morton Wheeler, 1865–1937." *National Academy of Sciences* 19:203–37.

Parker, Joan E. 2001. "Lydia Becker's 'School for Science': A Challenge to Domesticity." *Women's History Review* 10 (4):629–50.

Paul, Diane D. 1988. "The Selection of the 'Survival of the Fittest.'" *Journal of the History of Biology* 21 (3):411–24.

Philip, Kavita. 2004. *Civilizing Natures: Race, Resources, and Modernity in Colonial South India*. New Brunswick, NJ: Rutgers University Press.

Phillips, Louise, and Marianne W. Jorgensen. 2002. *Discourse Analysis as Theory and Method*. London: Sage.

Posey, Darrell A. 2002. *Kayapó Ethnoecology and Culture*, edited by Kristina Plenderleith. London: Routledge.

Prakash, Gyan. 1994. "Subaltern Studies as Postcolonial Criticism." *American Historical Review* 99 (5):1475–90.

———. 1999. *Another Reason: Science and the Imagination of Modern India*. Princeton, N.J.: Princeton University Press.

Pycraft W. P. 1913. *The Courtship of Animals*. London: Hutchinson.

Ransome, Hilda M. 1937. *The Sacred Bee in Ancient Times and Folklore*. Boston and New York: Houghton Mifflin.

Ratnieks, F. L. W., and T. Wenseleers. 2005. "Policing Insect Societies." *Science* 307:54–56.

Réaumur, René Antoine. [1743–44] 1926. *The Natural History of Ants*. Translated and annotated by William Morton Wheeler. New York: Knopf.

Reid, Roddey, and Sharon Traweek., eds. 2000. *Doing Science + Culture*. New York: Routledge.

Restivo, Sal. 1995. "The Theory Landscape in Science Studies." In *Handbook of Science and Technology Studies,* edited by Sheila Jasanoff, Gerald Markle, James C. Petersen, and Trevor Pinch, 95–113. Thousand Oaks, CA: Sage.

Reznikova, Zhanna. 2003. "Government and Nepotism in Social Insects: New Dimensions Provided by an Experimental Approach." *Euroasian Entomological Journal* 2 (1):3–14.

Richter, Maurice N., Jr. 1972. *Science as a Cultural Process*. Cambridge, MA: Schenkman.

Ritvo, Harriet. 1997a. *The Platypus and the Mermaid and Other Figments of the Classifying Imagination*. Cambridge: Harvard University Press.

———. 1997b. "Zoological Nomenclature and the Empire of Victorian Science." In *Victorian Science in Context,* edited by Bernard Lightman, 334–53. Chicago: University of Chicago Press.

———. 1999. "Border Trouble: Shifting the Line between People and Other Animals." In *Humans and Other Animals,* edited by Arien Mack, 67–86. Columbus: Ohio State University Press.

Robinson, Daniel N., ed. 1977. *Significant Contributions to the History of Psychology 1750–1920*. Vol. 3. Washington, DC: University Publications of America.

Root, A. I. [1877] 1966. *The ABC and XYZ of Bee Culture*. 33rd ed. Medina, Ohio: A. I. Root Co.

Ross, Herbert H. 1973. "Evolution and Phylogeny." In *History of Entomology,* edited by Ray F. Smith, Thomas E. Mittler, and Carroll N. Smith, 171–84. Palo Alto, CA: Annual Reviews.

Rossiter, Margaret W. 1982. *Women Scientists in America: Struggles and Strategies to 1940*. Baltimore: Johns Hopkins University Press.

Roth, Wolff-Michael. 2005. "Making Classifications (at) Work: Ordering Practices in Science." *Social Studies of Science* 35 (4):581–621.

Rowold, Katharina. 1996. *Gender and Science: Late Nineteenth-Century Debates on the Female Mind and Body.* Bristol, England: Thoemmes Press.

Rumbaugh, Duane. 1999. "Primitive Language and Cognition: Common Ground." In *Humans and Other Animals,* edited by Arien Mack, 301–20. Columbus: Ohio State University Press.

Sanders, Clinton R. 2003. "Actions Speak Louder Than Words. Close Relationships between Humans and Nonhuman Animals." *Symbolic Interaction* 26 (3):403–26.

Sapp, Jan. 1994. *Evolution by Association: A History of Symbiosis.* New York: Oxford University Press.

Schiebinger, Londa. 1989. *The Mind Has No Sex? Women in the Origins of Modern Science.* Cambridge: Harvard University Press.

Shapin, Steven. 1979. "Homo Phrenologicus: Anthropological Perspectives on an Historical Problem." In *Natural Order: Historical Studies of Scientific Culture,* edited by Barry Barnes and Steven Shapin, 41–67. London: Sage.

Sherman, Paul W., Eileen A. Lacey, Hudson K. Reeve, and Laurent Keller. 1995. "The Eusociality Continuum." *Behavioral Ecology* 6 (1):102–8.

Shouse, Ben. 2002. "Getting the Behavior of Social Insects to Compute." *Science* 295 (5564):2357.

Sismondo, Sergio. 2004. *An Introduction to Science and Technology Studies.* Malden, MA: Blackwell.

Sleigh, Charlotte. 2002. "Brave New Worlds: Trophallaxis and the Origin of Society in the Early Twentieth Century." *Journal of History of the Behavioral Sciences* 38(2):133–56.

———. 2003. *Ant.* London: Reaction Books.

Sommer, Marianne. 2000. *Foremost in Creation: Anthropomorphism and Anthropocentrism in National Geographic Articles on Non-Human Primates.* Bern, Switzerland: Peter Lang.

Sorenson, W. Connor. 1995. *Brethren of the Net: American Entomology 1840–1880.* Tuscaloosa: University of Alabama Press.

Spencer, Herbert. 1879. *The Data of Ethics.* New York: Appleton.

Strum, Shirley C., and Linda M. Fedigan, eds. 2000. *Primate Encounters: Models of Science, Gender and Society.* Chicago: University of Chicago Press.

Taylor, Peter J. 1988. "Technocratic Optimism, H. T. Odum, and the Partial Transformation of Ecological Metaphor after World War II." *Journal of the History of Biology* 21 (2):213–44.

Thompson, John. 1987. "Language and Ideology: A Framework for Analysis." *Sociological Review* 35 (3):516–36.

Thomson, Keith Stewart. 1998. "1798: Darwin and Malthus." *American Scientist* 86 (3): 226–29.

Todes, Daniel P. 1989. *Darwin without Malthus: The Struggle for Existence in Russian Evolutionary Thought.* New York: Oxford University Press.

Traweek, Sharon. 1988. *Beamtimes and Lifetimes: The World of High Energy Physicists.* Cambridge: Harvard University Press.

Tuchman, Gaye. 1994. "Historical Social Science: Methodologies, Methods, and Meanings." In *Handbook of Qualitative Research,* edited by Norman K. Denzin and Yvonna S. Lincoln, 306–23. Thousand Oaks, CA: Sage.

Tucker, Robert C., ed. 1978. *The Marx-Engels Reader.* New York: Norton.

von Frisch, Karl. [1927] 1953. *The Dancing Bees.* New York: Harvest.

Vorzimmer, Peter. 1969. "Darwin, Malthus, and the Theory of Natural Selection." *Journal of the History of Ideas* 30 (4): 527–42.

Wali, Mohan K. 1999. "Ecology Today: Beyond the Bounds of Science." *Nature and Resources* 35(2):38–50.

Wallace, Alfred Russel. 1872. "Wallace on the Origins of Insects." *Nature* 5: 350–51.

Ward, Lester. 1888. "Our Better Halves." *Forum* 6:266–76.

———. 1895. "Contributions to Social Philosophy II: Sociology and Cosmology." *American Journal of Sociology* 1 (2):132–45.

Wcislo, William T. 1997. "Are Behavioral Classifications Blinders to Studying Natural Variation?" In *The Evolution of Social Behavior in Insects and Arachnids,* edited by J. C. Choe and B. J. Crespi, 8–13. Cambridge: Cambridge University Press.

———. 2005. "Social Labels: We Should Emphasize Biology over Terminology and Not Vice Versa." *Annuals of Zoology Fennici* 42:565–68.

Weber, Max. 1978. *Economy and Society.* Edited by Guenther Roth and Claus Wittich. Berkeley and Los Angeles: University of California Press.

Weingart, Peter. 1994. "Biology as Social Theory: The Bifurcation of Social Biology and Sociology in Germany, circa 1900." In *Modernist Impulses in the Human Sciences 1870–1930,* edited by Dorothy Ross, 255–71. Baltimore: Johns Hopkins University Press.

West-Eberhard, Mary Jane. 1996. "Wasp Societies as Microcosms for the Study of Development and Evolution." In *Natural History and Evolution of an Animal Society: The Paper Wasp Case,* edited by Stephen Turillazzi and Mary Jane West-Eberhard, 290–317. Oxford: Oxford University Press.

———. 2003. *Developmental Plasticity and Evolution.* Oxford: Oxford University Press.

Wheeler, William Morton. 1910. *Ants: Their Structure, Development and Behavior.* New York: Columbia University Press.

———. 1923. *Social Life among the Insects.* New York: Harcourt, Brace.

———. 1928a. *The Social Insects: Their Origin and Evolution.* New York: Harcourt, Brace.

———. 1928b. *Foibles of Insects and Men.* New York: Knopf.
———. 1930. "Societal Evolution." In *Human Biology and Racial Welfare,* edited by Edmund V. Cowdry, 139–55. New York: Hoeber.
Williams, Vernon J. 1989. *From a Caste to a Minority: Changing Attitudes of American Sociologists toward African-Americans, 1896–1945.* New York: Greenwood Press.
Wilson, David Sloan, and Elliot Sober. 1989. "Reviving the Superorganism." *Journal of Theoretical Biology* 136:337–56.
Wilson, E. O. 1971. *The Insect Societies.* Cambridge: Harvard University Press.
———. 2003. "A United Biology: A Talk with E. O. Wilson." www.edge.org/3rd_culture/wilson03/wilson_index.html.
Withington, Anne Fairfax. 1988. "Republican Bees: The Political Economy of the Beehive in Eighteenth-Century America." *Studies in Eighteenth-Century Culture* 18:39–77.
Worster, Donald. 1985. *Nature's Economy: A History of Ecological Ideas.* Cambridge: Cambridge University Press.
Wright, Will. 1992. *Wild Knowledge: Science, Language and Social Life in a Fragile Environment.* Minneapolis: University of Minnesota Press.
Wyman, Leland C., and Flora L Bailey. 1964. *Navaho Indian Ethnoentomology.* University of New Mexico Publications in Anthropology Number 12. Albuquerque: University of New Mexico Press.
Young, Robert M. 1969. "Malthus and the Evolutionists: The Common Context of Biological and Social Theory." *Past and Present* 43:109–45.
———. 1985. *Darwin's Metaphor: Nature's Place in Victorian Culture.* Cambridge: Cambridge University Press.
Zeitlin, Irving M. 1997. *Ideology and the Development of Sociological Theory.* 6th ed. Upper Saddle River, NJ: Prentice Hall.
Zohar, Anat, and Shlomit Ginossar. 1998. "Lifting the Taboo Regarding Teleology and Anthropomorphism in Biology Education—Heretical Suggestions." *Science Education* 82 (6):679–97.

Index

Abraham, Itty, 29
actor network theory, science and, 31–32
Adams, C. C., 72, 78
Address, Political and Educational, 17
Aldrovandi, Ulysse, 6
Alger, Janet M. and Steven F., 95–96
Alic, Margaret, 52
Allee, Warder C., 174–75
   class and caste concepts of, 117–18
   ecology theory and, 74–78
   gender hierarchies and, 140, 157
   hierarchical analogies defended by, 174–75
   on legitimating terminology, 175
   matriarchal societies and, 160
Allen, Grant, 67–68
altruism, in eusocial insects, 12
American Journal of Sociology, 33, 74–75, 77, 116
American Social Insects, 82–83
analogy
   insect sociality research and, 94–96
   science's use of, 92–93
Animal Aggregations: A Study in General Sociology, 74–75
animal research, Chicago Ecology Group and, 74–76
Animal Societies: From the Bee to the Gorilla, 97
anthropomorphism
   research bias linked to, 93–114, 182–86
   social insect research and, 9–11
   sociology and, 89–90
ants
   anthropomorphism in study of, 9–11
   behavior and social research on, 6
   gender hierarchy with, 144–46
   Kayapó classification system for, 57–62
   language and reason in, 113–14
   matriarchal hierarchy and, 157–58
   military analogies in research on, 122–30
   Navajo classification system, 54–56
   slavery analogies in research on, 130–39
Ants, Bees and Wasps, 17
Ants, The, 6
Ants: Their Structure, Development and Behavior, 6
aphids
   eusocial classification of, 8
   generational overlap among, 13
Ardrey, Robert, 96, 124
Arluke, Arnold, 95
army ant, in social insect hierarchy, 11
artificial intelligence (AI), cultural biases in, 187

Bailey, Flora L., 54–56
Balas, M., 187
Barnes, Barry, 20
"Basic Comparisons of Human and Insect Societies," 76
Batra, Suzanne W., 5
Becker, Lydia E., 47–48
Beebe, William, 129
bees
   gender hierarchy with, 144–46
   individual and group behavior of, 108
   language of, 114
   racial analogies concerning, 134
Belt, Thomas, 127–29
Biological Symposia VIII, 77, 165
biology
   ecological theory and, 71–78
   feminist interpretations of, 161–62
   sociology and, 63–90, 94
"biracial organization" concept, 84
Bogardus, Emory, 85–86, 98, 100–101, 120–21
Butler, Charles, 44

Carpenter, Clarence Ray, 96–97
Caste and Ecology in the Social Insects, 179
castes and subcastes
  current challenges to analogies of, 179–80
  division of labor and, 12
  entomological-sociological connections concerning, 83–87
  eusociality and, 14–15
  hierarchical analogies of, 115–26
categories. *See* classification systems
cell research, comparative method and, 97
Chauvin, Remy, 97
Clark, J., 132, 141, 145
class
  current challenges to analogies of, 179–80, 184–86
  hierarchical analogies of, 115–26
  human sociality and, 15–19
  social organizational structure theory and, 3
classification systems
  current challenges to, 177–80
  indigenous science and, 53–62
  science studies and, 24–27
Clements, Frederic E., 64
Clutton-Brock, Juliet, 18
Cohen, I. B., 65
colonialism
  gender hierarchies and, 139–41
  hierarchical ranking and, 18–19, 126–27
  postcolonial science studies, 29–31
  racism and, 126–27
  slavery analogies shaped by, 134–39
color analogies, insect sociality research and, 133–34
communal behavior, subsocial insect classification, 5
companion animals, anthropomorphism concerning, 95–96
comparative method
  consciousness and, 102–7
  debates concerning, 97–111
  insect sociality and, 91–114

instinct and, 98–102
research possibilities of, 111–14
competition, evolution and role of, 69–71
Comstock, Anna B., 33, 45–46, 134, 168
concept transfer, entomology and sociology, 79–90
conflict, insect sociality and, 165–69
consciousness, insect sociality and, 102–7
continuum model of eusociality, 178–80
cooperation *vs.* competition
  evolution and, 68–71
  socialism of insects and, 163–69
Costa, James T., 8, 177–78
Cox, Oliver, 84
Creighton, William Steel, 55
crime, insect sociality and, 124
Crist, Eileen, 95–96, 113–14
critical discourse analysis
  entomology and, 2
  power relations and, 181–86
  scientific research and, 37–39
critical theory
  entomology and, 2
  methodology, 32–36
  science and, 23–24
  social-natural relationship and, 22–23
Cronin, Helena, 68–69, 155
cross-cultural factors
  insect classification systems and, 53–62
  superorganism concept and, 88–90
culture
  classification systems and, 57–63
  gender and, 52–53
Curley, Edwin A., 44–47

dairyman, in social insect hierarchy, 11
Dalke, Kate, 47
Darwin, Charles, 33, 64–68
  anthropomorphism in work of, 95–96
  comparative method and, 97
  concept transfer and theories of, 79–80
  on consciousness, 105
  entomology and, 65–66
  Kropotkin's critique of, 69–71

Index

primate-human relatedness and, 96
slavery analogies used by, 132–33
Davis, Natalie Zemon, 52–53
de animalibus insectis libri VII, 6
De Vore, Irven, 96
denaturalization, entomology and, 2
Descent of Man, The, 65–68
divine authority, science and concept of, 20–23
division of labor
  caste and, 12, 83–87, 115–26
  class and, 115–26
  current challenges to concepts of, 179–80
  Darwin's theories and, 80–81
  in eusocial insects, 4–5
  eusociality and, 14–15
  gender-based views of, 45–51
  human sociality and, 16–17
  indigenous insect classification systems and, 58–62
  race and, 129
  separate sphere analogy of, 47–48
  sociological-scientific interaction in theories of, 64
  superorganism concept and, 87–90
Division of Labor, The, 80–81
dominance hierarchies, 74–78. *See also* power relations
Douglas, Mary, 21
drone, in social insect hierarchy, 10, 144–46
Durkheim, Emile, 33, 79–81, 85, 100, 105–6

ecological theories
  in early twentieth-century, 71–78
  slavery analogies in, 132–39
  social and natural sciences and, 64
Eimer, Theodor, 34, 102, 110–11, 117, 156
Ellwood, Charles, 98–101, 105, 116
Elton, Charles, 78
Emerson, Alfred E., 76–78, 89, 98, 101, 112–13

Engels, Friedrich, 22, 42–43, 80–81
Entomological Society of London, 65–68
entomology
  anthropomorphism in, 94
  caste concepts in, 83–87
  class and caste concepts in, 117–26
  comparative method and, 98–111
  current debates on insect sociality, 177–80
  division of labor and, 80–83, 116–26
  ecological theory and, 71–78
  evolutionary theory and, 65–68
  gender issues in, 47–53
  instinct and, 101–2
  methodology of, 32–36
  queen imagery in, 44–47
  social construction of, 43–47
  sociology and, 15–19, 63–90, 181–86
Espinas, Alfred
  class and caste hierarchies and, 117–18, 164
  comparative debate and, 109
  ecological theory and, 74–75
  gender hierarchies and, 141, 144–45, 149–50
Essays on Population, 64
ethnobiology, indigenous science and, 53–62
eusociality
  categories of, 4–7
  class and caste and, 115–26
  current challenges to, 178–80
  definitions and terminology, 8–11, 13–15
  hierarchical analogies and, 116–26
  superorganism concept and, 11–13
evolutionary theory
  Allee's evolutionary comparison concept, 74–78
  comparative method debate and, 98
  entomology and, 65–68
  female force in, 160–63
  hierarchy of sociality and, 17–19
  inequality and, 23
  Kropotkin's ideas of, 68–71

evolutionary theory *(continued)*
  matriarchy and, 156
  nature theories and, 42–43
  nineteenth-century theories, 65–68
  organicist paradigm concerning, 34–36
  political implications of, 42–43
  postcolonial science studies and, 31
  scientific theory's intersection with, 32
  slavery analogies in, 132–39
  social *vs.* natural paradigms of, 65–68
  sociological view of, 64–90
Evrard, Eugene, 124, 134
exploitation, insect sociality and role of, 51–53

Fabre, Jean-Henri, 33, 124
farmer, in social insect hierarchy, 11
Feeling for the Organism, A, 28
female dominance in insect colonies, research discourse on, 140–54
Feminine Monarchie, 44
feminist science studies, 27–29, 182–87
  matriarchal insect societies, 155–63
Fichman, Martin, 159
Fielde, Adèle, 160–61
Fitzgerald, Terence D., 177–78
"folk science," indigenous science stereotyped as, 53–62
Forel, Auguste, 128
  comparative method and, 98, 101–2, 106, 110
  on cooperation among insects, 166–67
  eusociality and, 33
  feminist entomology and, 160–61
  on gender hierarchies, 141
  on human sociality, 167–68
  on insect behavior, 112, 124, 128–29
  on legitimating terminology, 175
  slavery analogies in research of, 130–32
  warfare analogies in research of, 165–66
Foster, William A., 8
Foucault, Michel, on classification systems, 25–26

Geddes, Patrick, 72
gender
  feminist science studies, 27–29, 187–88
  hierarchical analogies of, 139–54, 186
  indigenous hierarchies and, 58–62
  insect sociality and, 47–53
  matriarchal insect societies and, 156–63
  queen insect imagery and, 44–47
  scientific theory's intersection with, 32
  social organizational structure theory and, 3
generation overlap
  eusociality and, 12
  insect sociality and, 4
Genome News Network, 47
Gerard, R., 165
Giddings, Franklin Henry, 70–71
  comparative method debate and, 98
  on consciousness, 105
  eusociality and, 33
  human and insect sociality and, 15–16
  on instinct, 100
  natural sciences and, 42
  race and slavery analogies used by, 137–38
  sociological research and, 66–67
  superorganism and, 87–88
Gilman, Charlotte Perkins, 33, 42, 45, 49–50, 159–63
Goldenweiser, Alexander, 64–65
Gordon, Deborah, 2, 14–15, 87, 179–80
government, matriarchal hierarchies and, 158–63
Great Chain of Being worldview
  evolution and, 34
  hierarchical categories in, 17–18, 25–27
Gross, Neil, 64, 78
group, sociality and role of, 107–11
Group Mind, The, 103–4
guard bees, in social insect hierarchy, 10
Gurung, Astrid Björnsen, 60–61

Hamilton, William, 12, 89–90

Index 209

haplodiploidy, kin selection theory and, 12–13
Harding, Sandra, 36–37
Harvey, D., 21
Herbers, Joan, 177
heredity, caste concepts and, 84–87
Hess, David J., 22–23
Hesse, Mary B., 92
hierarchical analogies
  class/caste, 115–26
  current challenges to, 177–80
  gender, 139–54
  insect sociality and, 181–86
  lack of, in indigenous classification systems, 56–57
  legitimation of, 175–77
  Navajo insect classification system, 54–56
  postcolonial rejection of, 173–75
  of race, 126–39
  social organizational structure and, 2–3
Hölldobler, Bert, 6, 14
Holocaust, scientific research and impact of, 34–35
Honduran entomology, insect classification systems and, 59–62
honeybees, behavior and social research on, 5–6
Huber, François, 33, 121–22, 132, 176
"Human Ecology," 77
human sociality
  anthropomorphism in research on, 94–95
  class and caste and, 119–26
  consciousness and, 102–7
  cooperation vs. competition in, 69–71
  egalitarian structures and, 155
  indigenous hierarchies and, 58–62
  individual and group behavior and, 107–11
  insect sociality and, 15–19, 63, 91–92
  matriarchal hierarchies and, 158–59
  slavery and, 130–39

Huxley, Aldous, 73
Huxley, Julian, 77, 129, 140, 143, 152–53
Huxley, Thomas H., 168–69

inclusive fitness theory, 89–90
indigenous science
  colonialism and suppression of, 127, 172
  cross-cultural interpretations of insect sociality, 53–62
  social construction of, 41
individual, sociality and role of, 107–11
insect sociality
  anthropomorphism in research on, 94–95
  as behavioral model, 112–14
  class/caste hierarchies in, 83–87, 115–26
  comparative method and, 91–114
  consciousness and, 102–7
  cooperation vs. competition in, 69–71
  critical discourse analysis of, 37–39
  cross-cultural interpretations of, 53–62
  current debates on, 177–80
  Darwin's views on, 80–81
  definitions and terminology, 7–11
  division of labor and, 80–83
  ecological theory and, 72–78
  entomological definitions, 4
  feminist science analysis of, 28–29
  gender hierarchies in, 47–53, 139–54
  hierarchical analogies of, 115–26
  historical social construction of, 43–47
  indigenous classification systems and, 53–62
  individual and group behavior and, 107–11
  legitimating terminology, 175–77
  matriarchal insect societies, 155–63
  metaphorical language concerning, 92–93
  Navajo classification system, 54–56
  non-human interpretations of, 40–41
  postcolonial discourse on, 169–75
  power dynamics and, 22–23

insect sociality (continued)
  power relations and, 115–16
  racial hierarchy analogies and, 126–39
  research methodology concerning, 33–36
  in scientific literature, 1–2
  self-organizing models of, 40–41, 179–80
  slavery and, 130–39
  socialism and, 163–69
  social theory and human scale and, 15–19
  unnatural theory concerning, 41–43
instinct, comparative method and, 98–102
institutions, critical discourse analysis and role of, 38–39
Irvine, Leslie, 95–96
"Is There Any Specific Distinction between Male and Female Intellect?", 47

Jones, Greta, 64, 66

Kayapó Indians, insect classification system of, 53–54, 57–58
Keller, Evelyn Fox, 27–28
Kellogg, Vernon, 113, 130
Kelly, Kevin, 184
Kennedy, John S., 89, 95
kin selection theory, altruism and, 12–13
king, in social insect hierarchy, 10
Klamer, Arjo, 92
knowledge, social construction of, 40–41
Kropotkin, Peter, 33, 42, 65
  ecological theory and, 74
  evolution theory of, 68–71, 79, 81–83
  on insect sociality, 163–65, 174
Kuhn Thomas, 24, 35, 43

Laboratory Life, 24
Lacey, Eileen A., 178–80
Lakoff, George, 25–26, 91–92
Lamarckian evolutionary theory, 66–68
Land Nationalization doctrine, 67

language
  in insects, 113–14
  legitimating terminology for insect sociality, 175–77, 185–86
  sociological-entomological comparisons on, 106
Latour, Bruno, 24, 31–32
Latter, O. H., 141–43, 146–47
leadership, human sociality and, 16–17
Leonard, Thomas C., 92
Lewontin, R. C., 93
Life of the White Ant, The, 7
Limoges, Camille, 64–65, 79–80
Lubbock, John, 16–17, 33, 98, 105–6, 134–35
Lutz, Frank E., 48–53
Lyall, Malcolm, 156

Maeterlinck, Maurice
  comparative debate and, 106–7, 114
  gender hierarchies in research of, 141, 145–46, 148–50
  on insect sociality, 7, 33, 179–80
  slavery analogies in work of, 138
Malthus, Thomas
  Engels's critique of, 43
  insect sociality and, 23, 42–43
  population theory and, 64, 66, 68–71
Man-Made World; or, Our Androcentric Culture, The, 50, 163
Marais, Eugène N.
  class/caste analogies of, 123–24
  comparative method and, 104–5
  gender analogies of, 141, 143–44, 150–51
  on insect sociality, 7, 96–97
  on warfare, 165–66, 171–72
Margulis, Lynn, 93
marriage imagery, insect sociality research and, 49–53, 146–54
Martin, Emily, 49
Marx, Karl, 80–81, 122
matriarchal insect societies, 155–63
McClintock, Barbara, 28
McCook, Henry

Index

class/caste analogies of, 122–23, 125–27
gender analogies of, 145, 151
on insect sociality, 179–80
matriarchal hierarchy of, 157–59
racial analogies of, 133
on slavery, 170–71
on socialism in insect societies, 169
McDougall, William, 98, 103–7, 112
Mead, George Herbert, 33, 95–96, 98, 106–9, 112
Mendez de Torres, Luis, 44
Merian, Maria Sybilla, 52–53, 127, 172–73
Metamorphosis Insectorum Surinamensium, 52
metaphor, science's use of, 92–93
metaphoric language, science's use of, 92
Michener, Charles and Mary H., 4, 14, 82–83
military analogies, in insect sociality research, 122–29, 172–73
Milne-Edwards, Henri, 79–81
Mirowski, Phillip, 65
Morgan, C. Lloyd, 174–75
motherhood. *See also* matriarchal insect societies
  insect sociality and, 14, 146–54
  social constructions of, 46–47
Mukerjee, Radhakamal
  caste theory of, 83–86, 120–21, 124
  ecological theory and, 78–79
  evolutionary theory and, 72–73
  on gender hierarchies, 139–40
  on matriarchal insect societies, 156–58, 160
  postcolonialism and work of, 170
  social location theory and, 50–53
Mumford, Lewis, 72
Mutual Aid, 42, 70, 81
M-W paradigm, current challenges to, 177–78

Natural Images in Economic Thought: "Markets Read in Tooth and Claw," 65

natural selection
  caste concepts and, 84–87
  group vs. individual, 164
  inequality in, 23
  superorganisms and, 87–90
natural world
  ecological theories and, 71–78
  feminist science analysis of, 28–29
  insect sociality and, 42–43
  sociality and, 20–23
  sociological view of, 63–90
Navajo insect classification system, 54–55
neoanthropomorphism, 95–96
nonverbal communication, sociological-entomological comparisons on, 106
nurse, in social insect hierarchy, 10

Ogburn, William Fielding, 64–65
On the Origin and Metamorphoses of Insects, 17
oppositional discourse
  social constructionism and, 37
  social organizational structure and, 3
Organic Evolution, 34, 102
organicist paradigm
  evolutionary theory and, 34–36, 72–74
  matriarchal insect societies, 156–63
Origin of Civilization and the Primitive Conditions of Man, The, 17
Origin of Species, The, 64
Oster, George F., 179
Other Insect Societies: Reconsidering the Insect Sociality Paradigm, The, 8, 178

Pälawan indigenous insect classification system, 58–62
parental care, insect sociality and, 4, 13–14, 153–54
Pareto, Vilfredo, ecological theory and, 73–74
Park, Orlando, 76
Park, Robert E.
  Ecology Group and, 72, 74–78, 85

Park, Robert E. *(continued)*
  race and slavery analogies used by, 137
  sociology and, 66
Park, Thomas, 76
parthenogensis, generational overlap and, 13
Phemphigus obesinymphae, eusocial categorization of, 8
physical dimorphism, insect specialization and, 5
politics, scientific knowledge and, 24, 43
Posey, Darrel, 53–54, 57–62
postcolonial science studies, 29–31
  indigenous science and, 53–62
  insect sociality and, 169–75
power relations
  anthropomorphism and, 115
  insect sociality and, 22–23, 181–86
primate studies
  comparative debates in, 96–97
  feminist critiques of, 187
  language bias in, 93–94
primitive eusocial insects, classification of, 5
Principles of Animal Ecology, 76
Principles of Sociology, 87–88
Pycraft, W. P., 48–49, 141, 146

queen (insect)
  gender-based images of, 48–53, 140–54
  hierarchical category of, 9–10
  marriage and motherhood analogies in research on, 146–54
  matriarchal insect societies and, 157–63
  social construction of, 44–47

race
  caste and, 84–87
  hierarchical analogies of, 126–39, 184–86
  slavery analogies concerning, 130–39
  social organizational structure theory and, 3

rationality
  classification systems and, 26–27
  natural sciences and, 21–23
  reason, in insects, 113–14
Réaumur, René Antoine, 147–48, 152, 172, 176
Redfield, Robert, 77
Regional Sociology, 78
Reichard, Gladys A., 56
Reinheimer, Hermann, 156
reproductive division of labor, class and caste and, 115–26
Reuter, O. M., 176
Reznikova, Zhanna, 113
Romanes, George, 97
Ross, Edward A., 64
Ross, Herbert H., 66
Roth, Wolff-Michael, 25
Rumbaugh, Duane, 95

sacrifice, comparative debate concerning, 110–11
Sanders, Clinton R., 95
Schmidt, Karl, 76
science
  authority, sociality research and, 20–23
  classification systems and, 24–27
  critical theory concerning, 23–24
  feminist studies in, 27–29
  methodological approaches to, 32–36
  postcolonial studies, 29–31, 40–41, 169–75
  social construction of, 40–41
  sociology and, 63–90
  theological intersections in, 31–32
  war and peace and, 77–78
Science magazine, 96
Science of Life, The, 77
scout, in social insect hierarchy, 11
self-organizing models of insect sociality, 40–41, 179–80, 183–87
"Semantic Battles in a Conceptual War," 177–78

Shapin, Steven, 20
Sherman, Paul W., 178–80
Shouse, Ben, 183–86
situated knowledge, science and, 31–32
slave-maker, in social insect hierarchy, 10
slavery
    analogies in insect research, 177
    postcolonialist view of, 170
    racial analogies concerning, 130–39
    in social insect hierarchy, 10
Sleigh, Charlotte, 73–74, 126, 128
Smith, Adam, division of labor concept of, 15, 82, 184
social constructionism
    insect sociality and, 43–47, 139, 181–86
    knowledge and, 40–41
    scientific theory and, 36–37
social Darwinism
    colonialism and, 18–19
    natural selection, 23
social ecology, development of, 71–78
social evolution theory
    critical discourse analysis and, 38–39
    entomology and, 2–3
    race and slavery analogies in, 135–39
social insects. *See* insect sociality
Social Insects: Their Origin and Evolution, The, 88
socialism, insect sociality and, 163–69
sociality. *See also* human sociality; insect sociality
    authority and, 20–23
    cultural connotations of, 8–9
    definitions and terminology, 7–11
    hierarchical classification of, 5–7
"Social Life of the Baboon, The," 96
social location theory
    feminist science studies and, 28
    gender and reproduction and, 50–53
"Social Statics" ideology, 67
sociohistorical methods, scientific research and, 35–36
sociology and social science
    caste concepts and, 83–87, 124–26
    classification systems and, 26–27
    comparative method and, 98–111
    division of labor and, 80–83
    ecological theories and, 71–78
    entomology and, 15–19, 63–90, 181–86
    hierarchical analogies in, 116–26
    insect sociality and, 15–19
    natural sciences and, 20–23
    postcolonial science studies and, 30–31
    science and, 31–32, 63–90
    Spencer's influence in, 66–68
soldier, in social insect hierarchy, 10
Sommer, Marianne, 28–29, 147
Sorenson, W. Connor, 65, 71–72
Sorokin, Pitrim, 74
Soul of the Ape, 96
Soul of the White Ant, The, 7
specialized eusocial insects, classification of, 5
Spencer, Herbert, 42
    comparative method and, 97
    eusociality and, 11–12, 33
    evolutionary theory and, 65–68
    human sociality and, 16
    race and slavery analogies in work of, 136–37
    superorganism concept of, 87–88
    survival of the fittest concept of, 23, 42, 63
standpoint theory
    division of labor and, 48–53
    social constructionism and, 36–37
Structure of Scientific Revolutions, The, 24
subsocial insects
    classification of, 5
    definitions and terminology, 8–11
superorganisms
    comparative method and, 97
    entomological-sociological connections, 87–90
    eusocial insect colonies, 11–13
survival of the fittest

survival (continued)
   inequality in, 23
   insect sociality and, 42–43
   sociological paradigm and, 64

task-switching, insect sociality and, 15
Taylor, Walter, 72
technototemism, natural sciences and, 22–23
termites
   anthropomorphism in study of, 9–11
   behavioral and social research on, 7
   consciousness in, 104
   Tharu farmers, insect classification system of, 60–61
thief/robber, in insect hierarchy, 11

University of Chicago
   caste "school" at, 84–87
   Ecology Group, 64, 72, 74–78, 85, 157, 165
unnatural theory, insect sociality and, 41–43

von Frisch, Karl, 33, 113–14, 121, 123, 141, 146, 153

Wali, Mohan, 72–73
Wallace, Alfred Russel, 65–68, 136–37, 159
Wallace, Herbert Spencer, 67
war and peace, science and, 77–78
Ward, Lester, 101, 163
Warner, Lloyd, 84
Washburn, Sherwood, 96

wasps
   behavioral and social research on, 6–7
   Kayapó classification system for, 57–62
Weber, Max, 33, 98, 111–12
Wells, G. P., 77
Wells, H. G., 77
Western science
   indigenous science vs., 53–62
   insect classification and, 53–62
Wheeler, William Morton, 4, 6, 13–14, 33, 55
   caste concepts and, 85, 119–20, 125
   comparative method debate and, 98
   on cooperative behavior in insects, 165
   on division of labor, 116
   on gender hierarchies, 145, 156
   on legitimating terminology, 175–76
   organist ecology of, 34–36, 72–74, 77–78
   race and slavery analogies used by, 138
   on social cohesion, 82–83
   superorganism concept and, 88–90
Wilson, E. O., 4–6, 14, 82
women, in science, 27–29
Woolgar, Steve, 24
worker, in social insect hierarchy, 10
Worster, Donald, 34–35, 88
Wright, Will, 21–22
Wyman, Leland C., 54–56

Yerkes, Robert, 96–97
Young, Robert M., 31

Zeitlen, 30–31